T0140388

Springer Theses

Recognizing Outstanding Ph.D. Research

Aims and Scope

The series "Springer Theses" brings together a selection of the very best Ph.D. theses from around the world and across the physical sciences. Nominated and endorsed by two recognized specialists, each published volume has been selected for its scientific excellence and the high impact of its contents for the pertinent field of research. For greater accessibility to non-specialists, the published versions include an extended introduction, as well as a foreword by the student's supervisor explaining the special relevance of the work for the field. As a whole, the series will provide a valuable resource both for newcomers to the research fields described, and for other scientists seeking detailed background information on special questions. Finally, it provides an accredited documentation of the valuable contributions made by today's younger generation of scientists.

Theses are accepted into the series by invited nomination only and must fulfill all of the following criteria

- They must be written in good English.
- The topic should fall within the confines of Chemistry, Physics, Earth Sciences, Engineering and related interdisciplinary fields such as Materials, Nanoscience, Chemical Engineering, Complex Systems and Biophysics.
- The work reported in the thesis must represent a significant scientific advance.
- If the thesis includes previously published material, permission to reproduce this must be gained from the respective copyright holder.
- They must have been examined and passed during the 12 months prior to nomination.
- Each thesis should include a foreword by the supervisor outlining the significance of its content.
- The theses should have a clearly defined structure including an introduction accessible to scientists not expert in that particular field.

More information about this series at http://www.springer.com/series/8790

Abstract

This thesis investigates the resonant interaction between Rydberg atoms in a hot caesium vapour and terahertz frequency electromagnetic fields, and explores hyperfine quantum beats modified by driving an excited state transition in an inverted ladder scheme. The $21P_{3/2}$ caesium Rydberg atoms are excited using a three-step ladder scheme and we use a terahertz field resonant with the $21P_{3/2} \rightarrow 21S_{1/2}$ transition (0.634 THz), to measure Autler–Townes splitting of a 3-photon Rydberg electromagnetically induced transparency (EIT) feature. The Autler–Townes splitting allows us to infer the terahertz electric field amplitude, and we show a worked example measurement of a low-amplitude electric field, yielding 25 ± 5 mV cm^{-1}.

By driving an off-resonant Raman transition which combines the laser and terahertz fields, we restrict the Rydberg excitation to areas of the caesium vapour where the laser and terahertz fields spatially overlap. We show that the terahertz field intensity is proportional to the pixel intensity of a camera image of the atomic fluorescence, and demonstrate an image of a terahertz standing wave. The camera image is used to fit a model for a corresponding Autler–Townes spectrum, giving the scale of the electric field amplitude, and we use a video camera to record real-time images of the terahertz wave.

In the regime of intrinsic optical bistability, we study a Rydberg atom phase transition and critical slowing down, and we find that the terahertz field drives the collective Rydberg atom phase transition at low terahertz intensity ($I_T < 1$ Wm^{-2}). We measure a linear shift of the phase transition laser detuning with coefficient -179 ± 2 MHz W^{-1} m^2, and we use the frequency shift to detect incident terahertz radiation with sensitivity, $S \approx 90$ μW m^{-2} Hz$^{-1/2}$. When the system is initialised in one of two bistable states, a single 1 ms terahertz pulse with energy of order 10 fJ can permanently flip the system to the twin state.

sensitive detection of RF and microwave fields using Rydberg atoms at the University of Durham, University of Oklahoma, NIST and elsewhere, thus paving the way for the work described in these pages.

This thesis presents pioneering work on the development of electric field sensing and imaging in the terahertz range using thermal atomic vapour. The basic idea is to map THz photons which are typically very difficult to detect, on to optical photons which are easy to detect. The thesis also explores the phenomenon of optical bistability and Rydberg phase transitions, where the optical properties of the atomic system change abruptly in response to a small change in control parameters. Using this phenomenon, it is demonstrated that the application of very weak terahertz fields can lead to large changes in the system response.

The thesis begins with a brief overview of THz technologies in Chap. 1 before providing a theoretical underpinning of the work, describing the physics of Rydberg atoms and atom–light interactions in Chap. 2. Chapter 3 introduces and describes the experimental setup before Chap. 4 details an experimental and theoretical investigation into probing an atomic excited state transition using quantum beats. Chapter 5 discusses results on Rydberg phase transitions and optical bistability, and terahertz electrometry is introduced in Chap. 6. Chapter 7 contains a study of terahertz imaging and Chap. 8 explores terahertz-driven phase transitions. A summary and outlook is presented in Chap. 9.

The results contained in this thesis have stimulated significant interest within academia and industry. The work could pave the way for a new class of room temperature sensors capable of real-time, stand-off and in-situ measurements across the terahertz range and have impact across a wide range of applications.

Durham, UK   Kevin Weatherill
April 2018

Supervisor's Foreword

The terahertz frequency range, loosely defined as the $0.1 - 10 \times 10^{12}$ Hz range, has long been considered a very difficult part of the electromagnetic spectrum to work in. This is because terahertz radiation lies in the frequency gap between electronic, microwave technologies and infrared, photonic technologies—the so-called terahertz gap. Nevertheless, much progress has been made over the last few decades to harness the advantageous properties of terahertz radiation. For example, terahertz (THz) waves are used in security, medical and biological applications because they are low energy and non-ionising but also pass through many materials, such as clothing, plastics and paper, that are opaque in the visible region. In addition, since the ro-vibrational transitions of many molecules lie in the THz range, the spectroscopic distinction of materials such as explosives and drugs can be performed.

In parallel to advances in THz systems, technologies based upon thermal atomic vapour have been developing and can now offer high performance across a wide range of applications. For example, magnetic field sensors and gradiometers based on thermal vapour cells can provide state-of-the-art sensitivity and also have a much-reduced experimental overhead when compared to laser cooling or cryogenic experiments, allowing inexpensive and robust implementation. Sensors based upon atomic samples are particularly attractive because they are 'pre-calibrated', in the sense that each atom of the same isotope is identical and their well-known properties can be traced back to SI units. This is probably best exemplified in the atomic clock, which provides the SI standard for time, frequency and length. However, despite the successes of atomic magnetometers and clocks, by comparison, the measurement of electric fields using atomic vapours is not so well developed.

It has long been known that, due to the loosely bound electron, atoms in highly excited 'Rydberg' states have extreme sensitivity to electric fields. Rydberg atoms also have many electric dipole transitions in the microwave and THz range making them particularly sensitive to electric fields across these frequencies. However, it was only a decade ago that sensitive optical detection of Rydberg atoms was made possible through pioneering work on 'Rydberg EIT' at the University of Durham. This work was followed quickly by significant technical development in the

Christopher G. Wade

Terahertz Wave Detection and Imaging with a Hot Rydberg Vapour

Doctoral Thesis accepted by
Durham University, Durham, UK

 Springer

Author
Dr. Christopher G. Wade
Department of Physics
Durham University
Durham, UK

Supervisor
Dr. Kevin Weatherill
Department of Physics
Durham University
Durham, UK

ISSN 2190-5053   ISSN 2190-5061   (electronic)
Springer Theses
ISBN 978-3-030-06936-0   ISBN 978-3-319-94908-6   (eBook)
https://doi.org/10.1007/978-3-319-94908-6

Printed on acid-free paper

This Springer imprint is published by the registered company Springer Nature Switzerland AG
The registered company address is: Gewerbestrasse 11, 6330 Cham, Switzerland

Acknowledgements

I would like to express my heartfelt thanks to everyone who has helped me through the years of my Ph.D. study. Foremost, it has been a privilege to have the guidance of my supervisor, Kevin Weatherill—thank you for your patience, support and confidence. Thank you also to my second supervisor, Charles Adams, for your ideas and suggestions in our weekly meetings. Any of the work would have been impossible without a super team of people. Thank you to Massayuki, Natalia, Patrick, Hadrien and, in particular, Nick, to whom I am grateful for physical insight, laboratory guerrilla tactics and advice for life.

Good luck to Lucy, who will be continuing work on the project, and to Massayuki and Natalia who are embarking in new employment. I would also like to thank Mike Tarbutt, Claudio Balocco and Andrew Gallant for loaning us vital equipment and offering helpful advice. Thank you to Ifan Hughes for introducing me to Durham as a summer student.

I have been immensely happy working within the AtMol group—thank you to everyone who has made my time so enjoyable. In particular, thank you to Rob, James, Christoph, Dan, Pete, Mark and Danny for board games and Dani, Tommy and Nick for lunchtime runs. Thank you Alistair for all the ready-salted crisps(!), and the members of fAtMol for weekly cake and physics chat.

Finally, I would like to thank my family, Sally, Geof and David, for their support and for passing me their interest in science.

Contents

Chapter 1
Introduction

This thesis sits at the intersection of two distinct fields: Rydberg atomic physics and terahertz wave technology. Through the last century, experimental observations made in atomic physics have informed debate around topics ranging from the nature of quantum mechanics [1] to testing the equivalence principle [2], and from parity non-conservation [3] to Bose-Einstein condensation [4–6]. Each atom of the same isotope has identical, permanent properties and so atomic ensembles are ideal for conducting repeated and reproducible experiments. As a result, atomic physics has delivered a tool-kit for precision gravity [7], electromagnetic-field [8] and time [9] measurement.

While atomic physics has a long pedigree of fundamental enquiry and precision measurement, terahertz wave technology is a relatively nascent field, burgeoning from extensive development in recent decades [10]. Terahertz wave technology is concerned with producing, manipulating and detecting electromagnetic waves in the frequency range $0.1 \rightarrow 10\,\mathrm{THz}$. Science and technology each extend the reach of the other in a constructive cycle, and the study of terahertz fields is no exception. For example, the technological development of narrowband, tunable terahertz sources has revolutionised far-infrared spectroscopy [11], meanwhile scientific enquiry has lead to new materials [12] and metamaterials [13] suitable for terahertz technology. In this thesis we study the interaction of terahertz waves with atoms excited to high-lying Rydberg states, with a view towards technological applications.

1.1 Rydberg Atoms

Atoms excited to high-lying states with hydrogen-like energy levels are referred to as Rydberg atoms (see Sect. 2.1.1). Rydberg atoms have exaggerated properties and exhibit atomic radii, polarisabilities, lifetimes and dipole transition strengths that are several orders of magnitude greater than lower lying states. The properties of Rydberg atoms have been extensively studied and exploited for fundamental enquiries [14], and technological progress including optical processing [15, 16], quantum information processing [17, 18], quantum simulation [19], electrometry [20, 21] and cold-atom electron [22] and ion [23] beam sources.

© Springer International Publishing AG, part of Springer Nature 2018
C. G. Wade, *Terahertz Wave Detection and Imaging with a Hot Rydberg Vapour*,
Springer Theses, https://doi.org/10.1007/978-3-319-94908-6_1

Fig. 1.1 Caesium vapour
number density with
changing temperature,
calculated from [39]. The
dashed (full) line indicates
when the vapour is in
equilibrium with a solid
(liquid) phase

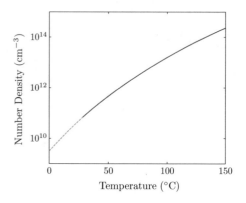

Rydberg atoms have been studied in very many different systems including atomic beams [24], thermionic diodes [25], ultra-cold ensembles [26], electrostatic traps [27], hollow-core fibres [28] and Bose-Einstein Condensates [29]. Even exotic Rydberg atoms comprising anti-matter [30] and excitons in condensed matter systems have been observed [31]. In this work, caesium Rydberg atoms are studied in a hot caesium vapour. Because hot atomic vapours are much easier to prepare than the systems mentioned above, they are of important technological significance. Their ease of use has allowed the development of practical magnetometry [32], miniaturised atomic clocks [33], magnetoencephalography [34], and microwave electrometry [20]. Work is under way to create efficient photon memories [35] and single photon sources [36]. Nevertheless, hot atomic vapours have also proved an effective medium for fundamental investigations concerning topics including atom-surface interactions [37] and atom-atom interactions [38].

An important virtue of atomic vapour is the tunabilty of the number density. Figure 1.1 shows how the number density of caesium vapour increases over four orders of magnitude as the vapour is heated from $0\,°C$ to $150\,°C$. With increased number density the average spacing between atoms decreases, and so inter-atomic interactions may be controlled through changing the temperature. Interactions between Rydberg atoms have been studied in thermal vapours including Van-der-Waals interactions [40], aggregation growth [41] and collective phase transitions [42, 43].

1.2 THz Technology

Over time sources of terahertz waves have become more powerful, and detectors more sensitive. There are now a great variety of bench-top techniques for producing terahertz fields [44, 45], including photomixers [46], quantum-cascade lasers [47] and frequency multiplication [48]. Pulsed terahertz fields can be generated using pulsed lasers incident on electro-optic crystals [49, 50] and photoconductive antennae [51].

The properties of terahertz radiation give terahertz technology wide ranging applications. Some materials that absorb optical light transmit THz radiation, and so THz technology can be suitable for security imaging, and detecting concealed objects in scenarios such as mail scanning and packaging inspection [52]. Astronomical sources of terahertz (Far-Infrared) reveal information about the universe [53], and because THz radiation is non-ionising it is an ideal candidate for medical imaging [54]. Although THz fields have the capacity for high bit-rate communication, atmospheric attenuation is a limitation for long-range, free-space transmission [55].

1.2.1 Terahertz Detectors

There are two types of terahertz detector: coherent detectors which measure both the amplitude and the phase of the THz field, and incoherent detectors which only measure the field intensity. Coherent detectors mix the incident field with a local oscillator, either another THz signal (heterodyne detection [56]) or a laser beam (photomixing [57]). Coherent detection of pulsed THz fields has led to the development of terahertz time-domain spectroscopy (THz-TDS), where the time-resolved electric field is Fourier transformed to deduce the spectral content of a THz pulse [58].

Incoherent detectors range from micro-fabricated single-photon counting devices [59], to micro-bolometer arrays [60]. Bolometers measure an incident THz field through a temperature change of the detector, which is read out using temperature-dependent resistance or a change of gas pressure (Golay cells [61]). Temperature-dependent resistors can be particularly sensitive when the resistor material is close to a phase transition, as in the case of transition edge sensors (TESs) [62]. Terahertz imaging can be performed using an array of pixel detectors, or by rastering the position of a single detector or probe. The former has an obvious speed advantage, though the latter can offer superior spatial resolution [63].

1.3 Terahertz Applications for Rydberg Atoms

Strong, electric-dipole transitions between Rydberg states give Rydberg atomic vapours a narrowband resonant response to terahertz frequency electric fields. Although any particular Rydberg state might only couple to a handful of resonant terahertz frequencies, transitions between different Rydberg states span the terahertz frequency range, any of which may be chosen for a terahertz device (Fig. 1.2). Matching atoms have identical transition frequencies, making Rydberg ensembles ideal candidates for frequency standards in the terahertz frequency range [65]. Furthermore the transition strengths are traceable to SI units, suggesting Rydberg atoms could be used as terahertz reference candles [66].

Rydberg atoms have previously been exploited as sources and detectors of terahertz waves. Early experiments demonstrated millimeter-wave Rydberg atom

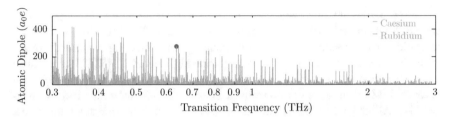

Fig. 1.2 Resonant transitions between Rydberg states in the THz band for rubidium and caesium. The transition used in this thesis is marked with a red dot. The transition frequencies and dipole strengths were calculated by Šibalić [64]

masers [67, 68], and terahertz imaging [69, 70]. The images were formed by ionising Rydberg atoms with a spatially-varying terahertz field, and focusing the resulting ions onto a detector. More recently electromagnetically induced transparency (EIT) in a hot vapour has been used for Rydberg microwave electrometry [20, 71]. Rydberg EIT maps Rydberg state dynamics onto a ground state optical transition [21], allowing a microwave or terahertz field to modify the transmission of a probe laser, which in turn provides a fast, optical read-out.

1.4 Thesis Overview

In this thesis we investigate the interaction of terahertz radiation with Rydberg atoms, culminating in the demonstration of calibrated real-time THz imaging (Chap. 7) and a room-temperature, THz-driven phase transition (Chap. 8). However, these results require some groundwork which is covered in the intervening chapters. In Chap. 2 we set out a theoretical foundation which forms the basis for interpreting later results using analytical results and computer simulations. We give an overview of caesium atomic energy levels and electric-dipole moments, and we introduce the master equation for describing atom dynamics. Chapter 3 outlines details of the experimental techniques used throughout the work, including the terahertz source and the laser system forming a three-step ladder excitation scheme. Although continuous-wave (CW) excitation was used for the majority of the work, measurements of atomic fluorescence after pulsed excitation lead to the observation of quantum beats, and in Chap. 4 we investigate how the quantum beats are modified by driving an excited state transition. In Chap. 5 we extend previous work investigating intrinsic Rydberg optical bistability and present new results including a spatial phase boundary and an experimental test of a recent theoretical model describing critical slowing down.

We commence our investigation of the interaction between Rydberg atoms and terahertz fields in Chap. 6, in which we describe the use the three step-ladder EIT scheme to perform Rydberg electrometry at 0.634 THz. In Chap. 7 we present a technique for real-time terahertz field imaging, and we calibrate the image sensitivity using Rydberg electrometry. The terahertz imaging technique allows the collection of images where terahertz fields are simultaneously propagating in opposite directions, and as a demonstration we present an image of a terahertz standing wave. Building on the work of Chap. 5, we report observations of a room-temperature phase transtion driven by a weak THz field, $I_T < 1\,\text{Wm}^{-2}$ (Chap. 8). The phase transition has the potential to be exploited as a sensitive THz detector, and we outline a pair of protocols to overcome hysteresis in the system response. In Chap. 9 we summarise the work, and give an outlook for future developments.

References

1. S. Haroche, J.-M. Raimond, *Exploring the Quantum: Atoms, Cavities and Photons* (Oxford, 2006)
2. S. Fray, C.A. Diez, T.W. Hänsch, M. Weitz, Atomic interferometer with amplitude gratings of light and its applications to atom based tests of the equivalence principle. Phys. Rev. Lett. **93**, 240404 (2004)
3. S.L. Gilbert, M.C. Noecker, R.N. Watts, C.E. Wieman, Measurement of parity nonconservation in atomic cesium. Phys. Rev. Lett. **55**, 2680 (1985)
4. K.B. Davis et al., Bose-Einstein condensation in a gas of sodium atoms. Phys. Rev. Lett. **75**, 3969 (1995)
5. C.C. Bradley, C.A. Sackett, J.J. Tollett, R.G. Hulet, Evidence of Bose-Einstein condensation in an atomic gas with attractive interactions. Phys. Rev. Lett. **75**, 1687 (1995)
6. M. Anderson, J. Ensher, M. Matthews, C. Wieman, E. Cornell, Observation of Bose-Einstein condensation in a dilute atomic vapor. Science **269**, 14 (1995)
7. M. de Angelis et al., Precision gravimetry with atomic sensors. Meas. Sci. Technol. **20**, 022001 (2009)
8. A. Facon et al., A sensitive electrometer based on a Rydberg atom in a Schrdinger-cat state. Nature **535**, 262 (2016)
9. H.S. Margolis, Frequency metrology and clocks. J. Phys. B **42**, 154017 (2009)
10. M. Tonouchi, Cutting-edge terahertz technology. Nat. Photonics **1**, 97 (2007)
11. P.U. Jepsen, D.G. Cooke, M. Koch, Terahertz spectroscopy and imaging-modern techniques and applications. Laser Photon. Rev. **5**, 124 (2011)
12. P. Tassin, T. Koschny, C.M. Soukoulis, Graphene for terahertz applications. Science **341**, 620 (2013)
13. B. Reinhard, O. Paul, M. Rahm, Metamaterial-based photonic devices for terahertz technology. IEEE J. Sel. Topics Quantum Electron. **19**, 8500912 (2013)
14. M. Brune et al., Quantum rabi oscillation: a direct test of field quantization in a cavity. Phys. Rev. Lett. **76**, 1800 (1996)
15. H. Gorniaczyk, C. Tresp, J. Schmidt, H. Fedder, S. Hofferberth, Single-photon transistor mediated by interstate Rydberg interactions. Phys. Rev. Lett. **113**, 053601 (2014)
16. D. Tiarks, S. Baur, K. Schneider, S. Dürr, G. Rempe, Single-photon transistor using a Förster resonance. Phys. Rev. Lett. **113**, 053602 (2014)
17. M. Saffman, T.G. Walker, K. Mølmer, Quantum information with Rydberg atoms. Rev. Mod. Phys. **82**, 2313 (2010)
18. D. Maxwell et al., Storage and control of optical photons using Rydberg polaritons. Phys. Rev. Lett. **110**, 103001 (2013)
19. H. Weimer, M. Muller, I. Lesanovsky, P. Zoller, H.P. Buchler, A Rydberg quantum simulator. Nat. Phys. **6**, 382 (2010)
20. J.A. Sedlacek et al., Microwave electrometry with Rydberg atoms in a vapour cell using bright atomic resonances. Nat. Phys. **8**, 819 (2012)

21. A.K. Mohapatra, T.R. Jackson, C.S. Adams, Coherent optical detection of highly excited Rydberg states using electromagnetically induced transparency. Phys. Rev. Lett. **98**, 113003 (2007)
22. A.J. McCulloch, D.V. Sheludko, M. Junker, R.E. Scholten, High-coherence picosecond electron bunches from cold atoms. Nat. Commun. **4**, 1692 (2013)
23. J.J. McClelland et al., Bright focused ion beam sources based on laser-cooled atoms. Appl. Phys. Rev. **3** (2016)
24. P. Goy, J.M. Raimond, G. Vitrant, S. Haroche, Millimeter-wave spectroscopy in cesium Rydberg states. Quantum defects, fine- and hyperfine-structure measurements. Phys. Rev. A **26**, 2733 (1982)
25. C.J. Sansonetti, C.J. Lorenzen, Doppler-free resonantly enhanced two-photon spectroscopy of np and nf Rydberg states in atomic cesium. Phys. Rev. A **30**, 1805 (1984)
26. J.-H. Choi, J.R. Guest, A.P. Povilus, E. Hansis, G. Raithel, Magnetic trapping of long-lived cold Rydberg atoms. Phys. Rev. Lett. **95**, 243001 (2005)
27. P. Lancuba, S.D. Hogan, Electrostatic trapping and in situ detection of Rydberg atoms above chip-based transmission lines. J. Phys. B **49**, 074006 (2016)
28. G. Epple et al., Rydberg atoms in hollow-core photonic crystal fibres. Nat. Commun. **5**, 4132 (2014)
29. R. Heidemann et al., Rydberg excitation of Bose-Einstein condensates. Phys. Rev. Lett. **100**, 033601 (2008)
30. A. Deller, A.M. Alonso, B.S. Cooper, S.D. Hogan, D.B. Cassidy, Electrostatically guided Rydberg positronium. Phys. Rev. Lett. **117**, 073202 (2016)
31. T. Kazimierczuk, D. Frohlich, S. Scheel, H. Stolz, M. Bayer, Giant Rydberg excitons in the copper oxide Cu_2O. Nature **514**, 343 (2014)
32. A. Wickenbrock, S. Jurgilas, A. Dow, L. Marmugi, F. Renzoni, Magnetic induction tomography using an all-optical ^{87}Rb atomic magnetometer. Opt. Lett. **39**, 6367 (2014)
33. S. Knappe et al., A microfabricated atomic clock. Appl. Phys. Lett. **85**, 1460 (2004)
34. T.H. Sander et al., Magnetoencephalography with a chip-scale atomic magnetometer. Biomed. Opt. Express **3**, 981 (2012)
35. D.J. Saunders et al., Cavity-enhanced room-temperature broadband Raman memory. Phys. Rev. Lett. **116**, 090501 (2016)
36. M.M. Müller et al., Room-temperature Rydberg single-photon source. Phys. Rev. A **87**, 053412 (2013)
37. A. Laliotis, T.P. de Silans, I. Maurin, M. Ducloy, D. Bloch, CasimirPolder interactions in the presence of thermally excited surface modes. Nat. Commun. **5**, 4364 (2014)
38. J. Keaveney et al., Cooperative lamb shift in an atomic vapor layer of nanometer thickness. Phys. Rev. Lett. **108**, 173601 (2012)
39. C.B. Alcock, V.P. Itkin, M.K. Horrigan, Vapour pressure equations for the metallic elements: 2982500K. Can. Metall. Q. **23**, 309 (1984)
40. T. Baluktsian, B. Huber, R. Löw, T. Pfau, Evidence for strong van der waals type Rydberg-Rydberg interaction in a thermal vapor. Phys. Rev. Lett. **110**, 123001 (2013)
41. A. Urvoy et al., Strongly correlated growth of Rydberg aggregates in a vapor cell. Phys. Rev. Lett. **114**, 203002 (2015)
42. C. Carr, R. Ritter, C.G. Wade, C.S. Adams, K.J. Weatherill, Nonequilibrium phase transition in a dilute Rydberg ensemble. Phys. Rev. Lett. **111**, 113901 (2013)
43. D. Weller, A. Urvoy, A. Rico, R. Löw, H. Kübler, Charge-induced optical bistability in thermal Rydberg vapor. Phys. Rev. A **94**, 063820 (2016)
44. C.M. O'Sullivan, J.A. Murphy, *Field Guide to Terahertz Sources, Detectors, and Optics* (SPIE PRESS, Bellingham, Washington USA, 2012)
45. R.A. Lewis, A review of terahertz sources. J. Phys. D **47**, 374001 (2014)
46. K.A. McIntosh et al., Terahertz photomixing with diode lasers in low-temperature-grown GaAs. Appl. Phys. Lett. **67**, 3844 (1995). https://doi.org/10.1063/1.115292
47. M.A. Belkin, F. Capasso, New frontiers in quantum cascade lasers: high performance room temperature terahertz sources. Phys. Scr. **90**, 118002 (2015)

48. A. Maestrini et al., A frequency-multiplied source with more than 1 mW of power across the 840–900 GHz band. IEEE Trans. Microwave Theory Tech. **58**, 1925 (2010)
49. M. Shalaby, C. Hauri, Demonstration of a low-frequency three-dimensional terahertz bullet with extreme brightness. Nat. Commun. **6**, 5976 (2015)
50. H. Hirori, A. Doi, F. Blanchard, K. Tanaka, Single-cycle terahertz pulses with amplitudes exceeding 1 MV/cm generated by optical rectification in LiNbO$_3$. Appl. Phys. Lett. **98**, 091106 (2011)
51. X. Ropagnol, F. Blanchard, T. Ozaki, M. Reid, Intense terahertz generation at low frequencies using an interdigitated ZnSe large aperture photoconductive antenna. Appl. Phys. Lett. **103**, 161108 (2013)
52. H.-B. Liu, H. Zhong, N. Karpowicz, Y. Chen, X.-C. Zhang, Terahertz spectroscopy and imaging for defense and security applications. Proc. IEEE **95**, 1514 (2007)
53. M. Griffin, Bolometers for far-infrared and submillimetre astronomy. Nucl. Instrum. Methods Phys. Res., Sect. A **444**, 397 (2000)
54. J.P. Guillet et al., Review of terahertz tomography techniques. J. Infrared Millim. Terahertz Waves **35**, 382 (2014)
55. H.-J. Song, T. Nagatsuma, Present and future of terahertz communications. IEEE Trans. Terahertz Sci. Technol. **1**, 256 (2011)
56. J. Gao et al., Terahertz heterodyne receiver based on a quantum cascade laser and a superconducting bolometer. Appl. Phys. Lett. **86**, 244104 (2005)
57. S. Verghese et al., Generation and detection of coherent terahertz waves using two photomixers. Appl. Phys. Lett. **73**, 3824 (1998)
58. M. Hangyo, M. Tani, T. Nagashima, Terahertz time-domain spectroscopy of solids: a review. Int. J. Infrared Millim. Waves **26**, 1661 (2005)
59. K. Ikushima, Single-photon counting and passive microscopy of terahertz radiation, in *Frontiers in Optical Methods* (Springer, 2014), pp. 197–212
60. A.W. Lee, Q. Hu, Real-time, continuous-wave terahertz imaging by use of a microbolometer focal-plane array. Opt. Lett. **30**, 2563 (2005)
61. P.S. Stefanova, J.M. Hammler, A.K. Klein, A.J. Gallant, C. Balocco, *Polymer-based micro-golay cells for THz detection*, in *2016 41st International Conference on Infrared, Millimeter, and Terahertz waves (IRMMW-THz)* (IEEE, 2016), pp. 1–2
62. M. Kehrt, J. Beyer, C. Monte, J. Hollandt, Design and characterization of a TES bolometer for Fourier transform spectroscopy in the THz range, in *Infrared, Millimeter, and Terahertz Waves (IRMMW-THz)* (IEEE, 2014), pp. 1–2
63. A.J. Huber, F. Keilmann, J. Wittborn, J. Aizpurua, R. Hillenbrand, Terahertz near-field nanoscopy of nanodevices. Nano Lett. **8**, 3766 (2008)
64. N. Šibalić, J. Pritchard, C. Adams, K. Weatherill, ARC: an open-source library for calculating properties of alkali Rydberg atoms (2016), arXiv:1612.05529
65. J. Raimond, P. Goy, G. Vitrant, S. Haroche, Millimeter-wave spectroscopy of cesium Rydberg states and possible applications to frequency metrology. J. Phys. Colloques (1981)
66. J.C. Camparo, Atomic stabilization of electromagnetic field strength using rabi resonances. Phys. Rev. Lett. **80**, 222 (1998)
67. L. Moi et al., Heterodyne detection of Rydberg atom maser emission. Opt. Commun. **33**, 47 (1980)
68. L. Moi et al., Rydberg-atom masers. I. A theoretical and experimental study of super-radiant systems in the millimeter-wave domain. Phys. Rev. A **27**, 2043 (1983)
69. M. Drabbels, L.D. Noordam, Infrared imaging camera based on a Rydberg atom photodetector. Appl. Phys. Lett. **74**, 1797 (1999)
70. A. Gurtler, A.S. Meijer, W.J. van der Zande, Imaging of terahertz radiation using a Rydberg atom photocathode. Appl. Phys. Lett. **83**, 222 (2003)
71. M.T. Simons et al., Using frequency detuning to improve the sensitivity of electric field measurements via electromagnetically induced transparency and Autler-Townes splitting in Rydberg atoms. Appl. Phys. Lett. **108** (2016)

Chapter 2
Atomic Structure and Atom-Light Interactions

All the work presented in this thesis was conducted using caesium vapour, and so we start by presenting an overview of caesium atomic structure and atom-light interactions. We discuss gross, fine and hyperfine structure, with specific reference to caesium, and use the master equation formalism to describe a 2-level atom interacting with a resonant electromagnetic field. We outline a method to infer spontaneous fluorescence into separate spatial modes using a decay operator. The theory presented in this chapter forms a foundation for interpreting the experimental results presented in the rest of the thesis.

Notation: We will write scaler quantities, a, in italics, and operators with hats, $\hat{a}$. We will write vectors, **a**, and vector operators, **â**, in bold.

2.1 Atomic Structure

The structure of atomic energy levels can be split according to a series of energy scales. The gross energy scale, the largest, is set by the ionisation energy of Hydrogen, $E_H \approx 13.6\,\text{eV}$, which is closely approximated by the Rydberg energy, $Ry \equiv \frac{1}{2}\alpha^2 m_e c^2 = 2\pi\hbar c R_\infty$, where m_e is the mass of an electron, c is the speed of light, $\alpha \simeq 1/137$ is the fine structure constant and R_∞ is the Rydberg constant [1]. Beyond the gross structure we consider the fine and hyperfine energy structures which have energy scales set by $\alpha^2 Ry$ and $\frac{m_e}{m_p}\alpha^2 Ry$ respectively, where m_p is the mass of a proton.

© Springer International Publishing AG, part of Springer Nature 2018
C. G. Wade, *Terahertz Wave Detection and Imaging with a Hot Rydberg Vapour*,
Springer Theses, https://doi.org/10.1007/978-3-319-94908-6_2

2.1.1   Gross Level Structure

The time-indepedent Schroedinger equation for a particle in a spherically symetrical potential, $V(r)$, is given by,

$$\left\{ \frac{-\hbar^2}{2m_e} \nabla^2 + V(r) \right\} \psi = E\psi, \tag{2.1}$$

where ψ is the particle wavefunction and E is the energy of the stationary state. Equation 2.1 may be separated into radial and angular components, and has solutions of the form,

$$\psi_{n,l,m_l} = R_{n,l}(r) Y_{l,m_l}(\theta, \phi), \tag{2.2}$$

where θ and ϕ are the polar and azimuthal angles respecively, Y_{l,m_l} are the spherical harmonics, and l and m_l correspond to the azimuthal and magnetic quantum numbers respectively. For the Coulomb potential between a proton and electron $V(r) \propto r^{-1}$, and the corresponding energies, $E_{n,l,m_l} = -Ry/n^2$ are degenerate in l and m_l and depend only on n. This expression gives the well known Rydberg formula that was used to describe early Hydrogen spectroscopic measurements. However, beyond the simple Schroedinger picture it is necessary to add relativistic and quantum electro-dynamic corrections to explain the full detail of Hydrogen atomic structure [2].

Alkali metal atoms, having a single valence electron, display energy level structure closely related to Hydrogen. However, the core electrons screen the extra nuclear charge and lead to a modified potential felt by the valence electron, $V^*(r)$. In general, imperfect screening of the nuclear charge leads to a lowering of the energy states and departure from a strict $1/r$ potential lifts the degeneracy of the l levels. The low angular momentum states have the greatest reduction in energy, because they have the highest valence electron probability density overlapping the inner core.

Far away from the nucleus, as $r \to \infty$, the potential $V^* \sim 1/r$, and it can be shown that for such potentials the energy levels go as,

$$E^*_{n,l,m_l} \simeq -\frac{1}{n^{*2}}, \tag{2.3}$$

where $n^* = n - \delta_l$ is the effective principal quantum number with δ_l as the quantum defect [2]. The quantum defect is different for each value of l, and in practice it also varies slighly with n [3]. Atoms excited to high n states have energies which adhere closely to Eq. 2.3, and are referred to as Rydberg atoms.

2.1.2   Fine Structure

As an electron moves inside an atom, it experiences an effective magnetic field as a consequence of the Lorentz transformation of the electric field from the nucleus and other electrons. The effective magnetic field interacts with the magnetic dipole of the electron spin, leading to an extra term in the Hamiltonian [2],

$$\hat{H}^{\text{spin}-\text{orbit}} = \frac{1}{2m_e^2 c^2} \frac{1}{r} \left(\partial_r V^*(r) \right) \hat{\mathbf{S}} \cdot \hat{\mathbf{L}}, \tag{2.4}$$

where $\hat{\mathbf{L}}$ is the vector orbital angular momentum operator for the electron, and $\hat{\mathbf{S}}$ is the vector spin operator for the electron. The spin-orbit term in the Hamiltonian mixes the states ψ_{n,l,m_l} into new eigenstates which are written,

$$|n, j, m_j, l, s\rangle = \sum C_{j,m_j,l,m_l,s,m_s} |n, l, m_l\rangle \otimes |s, m_s\rangle, \tag{2.5}$$

where $|n, l, m_l\rangle \otimes |s, m_s\rangle$ is an electronic state with spatial wavefunction ψ_{n,l,m_l} and spin state $|s, m_s\rangle$ and C_{j,m_j,l,m_l,s,m_s} are Clebsch-Gordon coefficients. We now use the total angular momentum,

$$\hat{\mathbf{J}}^2 = \left(\hat{\mathbf{L}} + \hat{\mathbf{S}} \right)^2,$$

which has eigenvalues $j(j+1)\hbar^2$ and the azimuthal angular momentum, $\hat{j}_z = \hat{l}_z + \hat{s}_z$, which has eigenvalues m_j, to label the states. Because $\hat{\mathbf{J}}^2$ and $\hat{j}_z$ both commute with $\hat{H}^{\text{spin}-\text{orbit}}$, j and m_j are good quantum numbers. The change in energy of the new eigenstates is proportional to,

$$\langle \hat{\mathbf{S}}.\hat{\mathbf{L}} \rangle = \frac{1}{2} \left(j(j+1) - l(l+1) - s(s+1) \right),$$

and we can see that the energies of states with $l = 0$ are unchanged as in this case $j = s$ [2]. Single valence electron atoms such as caesium have spin $s = 1/2$ so states with $l > 0$ are split into pairs of degenerate manifolds $j = l + 1/2$ and $j = l - 1/2$. The energy separation of the states is smaller for wavefunctions with larger n or l because the valence electron spends more time away from the nucleus where the electric field, and therefore the effective magnetic field, is smaller.

2.1.3 Hyperfine Structure

The electrons of an atom create a magnetic field which is felt by the nucleus. The nucleus has a magnetic moment, $\boldsymbol{\mu}_I = g_I \mu_N \mathbf{I}$, where $\mu_N = e\hbar/2m_P$ is the nuclear magneton, $\mathbf{I}$ is the nuclear spin and g_I is nuclear gyromagnetic ratio which depends on the internal structure of the nucleus. The leading term in the interaction between the nuclear magnetic moment and the magnetic field gives the hyperfine interaction,

$$\hat{H}^{\text{Hyperfine}} = A \hat{\mathbf{I}} \cdot \hat{\mathbf{J}}, \tag{2.6}$$

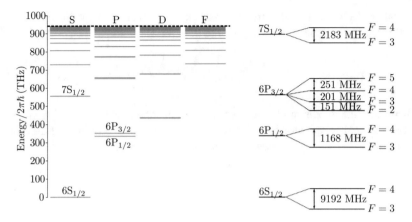

Fig. 2.1 Caesium energy levels [7]. Left: S,P,D and F energy levels from the ground state to the ionisation threshold (dashed line). Right: Schematic representation of the hyperfine splitting of selected states

where the magnetic dipole constant, A, depends on the nuclear magnetic moment and the magnetic field from spin and orbital motion of the electron projected onto $\mathbf{J}$ [4]. Once again the interaction mixes the atomic states, and the new eigenstates are given by,

$$|n, f, m_f, j, l, s\rangle = \sum C_{f,m_f,j,m_j,I,m_I} |n, j, m_j, l, s\rangle \otimes |I, m_I\rangle, \qquad (2.7)$$

where $|I, m_I\rangle$ is the spin state of the nucleus and the new states have energy given by,

$$A\langle \hat{\mathbf{I}} \cdot \hat{\mathbf{J}} \rangle = \frac{A}{2} \left(F(F+1) - J(J+1) - I(I+1) \right),$$

and for caesium, magnetic dipole constants have been measured [5, 6],

$$A_{6P_{3/2}} = h.50.275(3) \text{ MHz, and } A_{6P_{1/2}} = h.291.89(8) \text{ MHz}.$$

The hyperfine energy structure of caesium Rydberg states is too small to be resolved by the experimental methods used in this thesis.

2.1.4   Caesium Atomic Structure

We conclude this section by outlining the fine and hyperfine structure of caesium (Fig. 2.1). On the left we show caesium energy levels from the ground state, $6S_{1/2}$, to the ionisation limit and up to orbital angular momentum quantum number $l = 3$. Throughout the work in this thesis we use a ladder excitation scheme to excite high lying Rydberg levels shown ($E/(2\pi\hbar) > 850$ THz). On the right we highlight the

$6S_{1/2}$, $6P_{3/2}$ and $7S_{1/2}$ states which form the 'rungs' of the ladder, and we label the hyperfine splitting.

2.2 Atom Light Interaction

So far we have only considered the Hamiltonian of an isolated atom. To describe an atom interacting with electromagnetic fields it is necessary to construct a Hamiltonian that not only includes the internal energy of the atom, but also the energy of the field and the coupling between the two. For a two-level atom with ground state $|g\rangle$ and excited state $|e\rangle$ separated by energy $\hbar\omega_0$, we make the rotating wave approximation and obtain [8],

$$\hat{H} = \sum_{k,\epsilon} \hbar v_k \left(\hat{a}^{\dagger}_{k,\epsilon} \hat{a}_{k,\epsilon} + \frac{1}{2} \right) + \hbar\omega_0 |e\rangle\langle e| + \sum_{k,\epsilon} \hbar g_{k,\epsilon} \left(|e\rangle\langle g|\hat{a}_{k,\epsilon} + |g\rangle\langle e|\hat{a}^{\dagger}_{k,\epsilon} \right),$$

(2.8)

where the transition between states $|g\rangle$ and $|e\rangle$ is coupled to field modes with wavevector k, polarisation ϵ and frequency v_k by the coupling factor $g_{k,\epsilon}$, and the creation and annihilation operators for the field modes are $\hat{a}^{\dagger}_{k,\epsilon}$ and $\hat{a}_{k,\epsilon}$ respectively.

2.2.1 Dipole Matrix Elements

When the transition between two atomic states is electric-dipole allowed, the biggest contribution to $g_{k,\epsilon}$ comes from electric dipole coupling. In this section we find an expression for the electric dipole matrix element between two atomic states. We make the dipole approximation, and assume that the spatial extent of the atom is very much smaller than the wavelength of the resonant field mode, and we write [8],

$$g_{k,\epsilon} = \epsilon \cdot \mathbf{d}_{\text{eg}} \mathscr{E}_k e/\hbar,$$

where ϵ is the polarisation of the electric field, e is the charge on an electron, and $\mathbf{d}_{\text{e,g}} = \langle g|\hat{\mathbf{r}}|e\rangle$ is the dipole matrix element between states $|g\rangle$ and $|e\rangle$, with $\hat{\mathbf{r}}$ the vector position operator for the valence electron with respect to the nucleus. The factor $\mathscr{E}_{k,\epsilon} = \sqrt{\hbar v_k/2\epsilon_0 V}$ has dimensions of electric field amplitude, and depends on the volume of the field modes, V. Because the coupling is set by the spatial electron wavefunction, it is convenient to write atomic states in the $|n, l, m_l\rangle \otimes |s, m_s\rangle \otimes |I, m_I\rangle$ basis, even though these states are not eigenstates of the atomic Hamiltonian. In this basis a transition is dipole-allowed if the two states have identical nuclear and spin wavefunctions, and the dipole $\mathbf{d}_{n',l',m',n,l,m} = \langle n, l, m_l|\hat{\mathbf{r}}|n', l', m'_l\rangle \neq 0$, a condition which is fulfilled if $|l' - l| = 1$ and $m'_l - m_l = \{-1, 0, 1\}$. The dipole, $\mathbf{d}_{\text{e,g}}$, between the two eigenstates of the atomic Hamiltonian is then a coherent sum of the component dipoles $\mathbf{d}_{n',l',m',n,l,m}$, weighted by the decomposition of $|e\rangle$ and $|g\rangle$. In order to find $\mathbf{d}_{n',l',m',n,l,m}$ we work in the convenient basis of vectors,

$$\boldsymbol{\mu}_{-1} = \frac{1}{\sqrt{2}}(\mathbf{e}_x - i\mathbf{e}_y), \tag{2.9}$$

$$\boldsymbol{\mu}_0 = \mathbf{e}_z, \tag{2.10}$$

$$\boldsymbol{\mu}_{+1} = \frac{1}{\sqrt{2}}(\mathbf{e}_x + i\mathbf{e}_y), \tag{2.11}$$

giving,

$$\boldsymbol{\epsilon} \cdot \mathbf{d}_{n,l,m,n',l',m'} = \sum_q (\boldsymbol{\mu}_q \cdot \boldsymbol{\epsilon}) \langle n, l, m_l | \mu_q | n', l', m_l' \rangle, \tag{2.12}$$

where $\mu_q = \boldsymbol{\mu}_q \cdot \hat{\boldsymbol{r}}$. Noting that $\mu_q = \sqrt{\frac{4\pi}{3}} r Y_{1,q}(\theta, \phi)$, where $\{r, \theta, \phi\}$ are the coordinates of the valence electron, we use the general wavefunction given in Eq. 2.2 to obtain,

$$\langle n, l, m_l | \mu_q | n', l', m_l' \rangle = \sqrt{\frac{4\pi}{3}} \iint Y_{1,q} Y_{l,m}^* Y_{l',m'} \sin\theta \, d\theta \, d\phi \int_0^\infty r R_{n,l}^* R_{n',l'}' r^2 dr, \tag{2.13}$$

where we have separated the integral into angular and radial parts. This expression can be written [9],

$$\langle n, l, m_l | \mu_q | n', l', m_l' \rangle = (-1)^{l'+1-m_l} \begin{pmatrix} l' & 1 & l \\ m_{l'} & q & -m_l \end{pmatrix} \langle n, l || \mu || n', l' \rangle, \tag{2.14}$$

where $(\ldots)$ denotes a Wigner 3-j coefficient and we have used the reduced dipole matrix element [9],

$$\langle n, l || \mu || n', l' \rangle = (-1)^l \sqrt{(2l+1)(2l'+1)} \begin{pmatrix} l' & 1 & l \\ 0 & 0 & 0 \end{pmatrix} \int_0^\infty r R_{n,l}^* R_{n',l'}' r^2 dr, \tag{2.15}$$

which has no angular dependence.

To find the sum of the matrix elements $\mathbf{d}_{n,l,m,n',l',m'}$ corresponding to $\mathbf{d}_{eg}$, we write the fine structure reduced dipole matrix element,

$$\langle j || \mu || j' \rangle = (-1)^{l+s+j'+1} \sqrt{(2j+1)(2j'+1)} \begin{Bmatrix} j & 1 & j' \\ l' & s & l \end{Bmatrix} \langle n, l || \mu || n', l' \rangle \tag{2.16}$$

where $\{\ldots\}$ denotes a Wigner 6-j symbol. For the hyperfine basis we write,

$$\langle F || \mu || F' \rangle = (-1)^{j+I+F'+1} \sqrt{(2F+1)(2F'+1)} \begin{Bmatrix} F & 1 & F' \\ j' & I & j \end{Bmatrix} \langle j || \mu || j \rangle \tag{2.17}$$

Finally we write the dipole matrix element between two hyperfine states,

$$\langle F, m_F | \mu_q | F', m'_F \rangle = (-1)^{F'+1-m_F} \begin{pmatrix} F' & 1 & F \\ m_{F'} & q & -m_F \end{pmatrix} \langle F || \mu || F' \rangle. \qquad (2.18)$$

We use the formalism laid out here to interpret hyperfine quantum beats in Chap. 4 and the THz electric field amplitude in Chap. 6.

2.2.2 Master Equation

Solving the full quantum description of an atom interacting with an electric field is difficult and unnecessary in a lot of situations. Instead we treat the atom as a open quantum system interacting with an environment. For a laser-driven, two-level atom, the open quantum system comprises the atom interacting with the (classical) electric field of the laser, and the environment corresponds to the vacuum modes which are responsible for spontaneous decay. We write the Rabi frequency,

$$\Omega = \epsilon \cdot \mathbf{d}_{eg} E / \hbar, \qquad (2.19)$$

where E is the electric field amplitude at the position of the atom, giving the Hamiltonian for the atom in the dressed state picture as,

$$\hat{H}^{2-\text{level}} = \frac{\hbar}{2} \begin{pmatrix} 0 & \Omega \\ \Omega & -2\Delta \end{pmatrix}, \qquad (2.20)$$

where $\Delta = \omega - \omega_0$ is the detuning of the laser from the atomic transition frequency, ω_0, and we have used the dressed state basis.

We use the master equation and density matrix formalism to describe the system [10]. The density matrix, ρ, is a generalisation of the quantum state vector which allows us to describe the quantum system in a mixed state involving a probabilistic mixture of quantum states. The Lindblad master equation describes the system dynamics,

$$\frac{\partial \rho}{\partial t} = -\frac{i}{\hbar} [\hat{H}, \rho] + \hat{\mathcal{L}}\{\rho\}, \qquad (2.21)$$

where the decay super-operator has the general form,

$$\hat{\mathcal{L}}\{\rho\} = \sum_j C_j \rho C_j^\dagger - \frac{1}{2} \left(\rho C_j^\dagger C_j + C_j^\dagger C_j \rho \right),$$

where C_j are the collapse operators for the system, representing interaction with the environment. For spontaneous decay from the excited state to the ground state we use,

$$C_e = \sqrt{\Gamma_e} |g\rangle \langle e|,$$

where Γ_e is the lifetime of the excited state. Equation 2.21 is a set of coupled 1st order linear differential equations, which may be solved analytically. It is simple to generalise the treatment to describe multi-level atoms, although numerical integration is required to solve the equations.

2.2.3   Spontaneous Decay

Despite treating the atom as an open quantum system and neglecting the vacuum field modes, it is still possible to infer some properties of the atomic fluorescence. A textbook treatment of the fluorescence from a two-level atom is given by Scully and Zhubairy [8] and the fluorescence intensity at an observation point is calculated by integrating over all possible field modes. It is shown that a field mode only makes a contribution if it is both coupled to the atomic dipole and also non-vanishing at the observation point in the far field. These geometric considerations help us to arrive at the simple result for the expected instantaneous fluorescence intensity at the position of a sensor, $\mathbf{s}$, from an atom in the pure state $|\psi\rangle = \alpha_g |g\rangle + \alpha_e |e\rangle$,

$$I^{2\text{-level}}(\mathbf{s}, t, \boldsymbol{\epsilon}_s) = \frac{k_0^4}{(4\pi \epsilon_0 s)^2} \left| \alpha_e(t) e^{-i\omega_0 t} \boldsymbol{\epsilon}_s \cdot \mathbf{d}_{e,g} \right|^2, \qquad (2.22)$$

where $k_0 = \omega_0 / c$ and $\boldsymbol{\epsilon}_s$ is the polarisation to which the detector is sensitive (we note that the retardation time, s/c, has been neglected). When the atoms are in a superposition of excited states the situation becomes more complicated. The contribution from each of the excited states must be added coherently, taking phase information into account. For an atom with several excited states, $|e_i\rangle$, and ground states, $|g_k\rangle$, we write instead,

$$I(\mathbf{s}, t, \boldsymbol{\epsilon}_s) = \sum_k \frac{k_0^4}{(4\pi \epsilon_0 s)^2} \left| \sum_i \alpha_i(t) e^{-i\omega_{i,k} t} \boldsymbol{\epsilon}_s \cdot \mathbf{d}_{i,k} \right|^2, \qquad (2.23)$$

where $\hbar \omega_{i,k}$ is the energy interval between stated $|g_k\rangle$ and $|e_i\rangle$. We note that the expression, $\boldsymbol{\epsilon}_s \cdot \mathbf{d}_{i,k}$, has the same form as Eq. 2.12. In the hyperfine basis we write,

$$\boldsymbol{\epsilon}_s \cdot \mathbf{d}_{F,m_f,F',m'_f} = \sum_q \left(\boldsymbol{\mu}_q \cdot \boldsymbol{\epsilon}_s \right) \langle F, m_F | \mu_q | F', m'_F \rangle. \qquad (2.24)$$

We expand Eq. 2.23 to give the expression given by Haroche [11],

$$I(\mathbf{s}, t, \boldsymbol{\epsilon}_s) = \sum_k \frac{k_0^4}{(4\pi\epsilon_0 s)^2} \sum_{i,j} \alpha_i(t)\alpha_j^*(t) e^{-i(\omega_{i,k}-\omega_{j,k})t} \langle e_i|\hat{\mathbf{r}}\cdot\boldsymbol{\epsilon}_s|g_k\rangle\langle g_k|\hat{\mathbf{r}}\cdot\boldsymbol{\epsilon}_s|e_j\rangle,$$

(2.25)

where $\hat{\mathbf{r}}$ is the position operator for the valence electron with respect to the nucleus.

In the density matrix formalism, an atom prepared in a pure state which undergoes spontaneous decay evolves into a mixed state. Using the density matrix, $\rho(t)$, we identify Eq. 2.25 equivalent to [11],

$$I(s, t, \boldsymbol{\epsilon}_s) = \mathrm{Tr}(\rho(t)\mathscr{L}_{\boldsymbol{\epsilon}_s})$$

(2.26)

where the decay operator,

$$\mathscr{L}_{\boldsymbol{\epsilon}_r} = \sum_k \frac{k_0^4}{(4\pi\epsilon_0 s)^2} \left(\hat{\mathbf{r}}\cdot\boldsymbol{\epsilon}_s|g_k\rangle\langle g_k|\hat{\mathbf{r}}\cdot\boldsymbol{\epsilon}_s\right),$$

(2.27)

gives the atomic fluoresence. In Chap. 4 we will use the master equation and decay operator to model quantum beats.

2.3 Conclusion

The theory outlined in this chapter will support the work presented throughout the rest of the thesis. In Chap. 4 hyperfine beats in D_2 fluorescence are analysed and used to read out dynamics of the transition driven by the coupling laser. The analysis of the data requires a comprehensive understanding of the hyperfine structure, optical Bloch equations and the spontaneous emission formalism outlined in Sect. 2.2.3. In Chap. 5 inter-atomic interactions lead to non-linear dynamics, and we extend the master equation formalism to make a phenomenological 2-level model, exhibiting optical bistability. The fourth driving field (THz) is introduced in Chap. 6, and we use a 5-level master equation treatment to simulate Autler-Townes splitting for terahertz electrometry.

References

1. P.J. Mohr, B.N. Taylor, D.B. Newell, CODATA recommended values of the fundamental physical constants: 2006. Rev. Mod. Phys. **80**, 633 (2008)
2. C.J. Foot, *Atomic Physics* (Oxford University Press, 2005)
3. T.F. Gallagher, *Rydberg Atoms* (Cambridge University Press, 1994)
4. E. Arimondo, M. Inguscio, P. Violino, Experimental determinations of the hyperfine structure in the alkali atoms. Rev. Mod. Phys. **49** (1977)
5. E. Tanner, C. Wieman, Precision measurement of the hyperfine structure of the ^{133}Cs $6P_{3/2}$ state. Phys. Rev. A **38**, 1616 (1988)

6. R.J. Rafac, C.E. Tanner, Measurement of the ^{133}Cs $6p^2 P_{1/2}$ state hyperfine structure. Phys. Rev. A **56**, 1027 (1997)
7. J. Sansonetti, Accurate energies of nS, nP, nD, nF, and nG levels of neutral cesium. Phys. Rev. A **35**, 4650 (1987)
8. M.O. Scully, M.S. Zubairy, *Quantum Optics* (1999)
9. I.I. Sobelman, *Atomic Spectra and Radiative Transitions*, vol. 12 (Springer Science & Business Media, 2012)
10. K. Blum, *Density Matrix Theory and Applications*, vol. 64 (Springer Science & Business Media, 2012)
11. S. Haroche, *Quantum Beats and Time-Resolved Fluorescence Spectroscopy*, vol. 13 of *Topics in Applied Physics*, Chap. 7 (Springer-Verlag, 1976), p. 253

Chapter 3
Experimental Methods

We describe a set of experimental techniques which are shared by the experiments presented in this thesis. Much of the work uses a three-step laser excitation scheme to excite caesium Rydberg atoms, which are then manipulated by a terahertz (THz) frequency field, and monitored through laser transmission and atomic fluorescence. Our requirements therefore include a temperature stabilised caesium vapour, a frequency stabilised laser system, a THz source and equipment to measure laser beam and fluorescence intensity. We address each element in turn, giving details of the system that was used. In subsequent chapters we will note deviations from, or additions to the methods described here.

3.1 Caesium Vapour

All the experiments described in this thesis were conducted using a caesium vapour contained within a 2 mm long quartz glass cell at temperatures between 19 °C and 90 °C. In Sect. 1.1 we described some of the advantages of using alkali-metal atomic vapour, and we noted that the strong temperature dependence of the vapour pressure permits the tuning of the number density, and hence the inter-atomic separation. The inter-atomic separation is an important parameter for phenomena which depend upon interactions between atoms. Therefore, the combination of the atomic interaction strength and probe laser absorption cross section set the scale for the 2 mm optical path length through the cell, as the optical depth must be sufficient (though not excessive) when the number density is suitable for observing the atomic interactions investigated in this thesis.

While the strong temperature dependence allows easy access to a wide range of atom number density, it also means that the cell temperature must be carefully controlled to avoid unwanted changes. We heat the cell using either of two purpose built heaters, designed by Weller and Kondo. The first heater works as an oven: the glass cell is encased in non-magnetic steel which is itself heated by ceramic resistors. Holes drilled through the steel allow optical access for laser beams passing through

© Springer International Publishing AG, part of Springer Nature 2018   19
C. G. Wade, *Terahertz Wave Detection and Imaging with a Hot Rydberg Vapour*,
Springer Theses, https://doi.org/10.1007/978-3-319-94908-6_3

Fig. 3.1 Weak probe
transmission profile of the
D_2 $F = 4 \rightarrow F' = \{3, 4, 5\}$
caesium transition in a 2 mm
cell at 48 °C (blue),
68 °C (yellow) and
77 °C (red). The
temperatures are inferred
from a model fit (black lines)
in which temperature is a
free parameter. The relative
laser detuning was measured
using an etalon, and a global
frequency shift was used as a
fit parameter, setting the
$F = 4 \rightarrow F' = 5$ transition
to zero

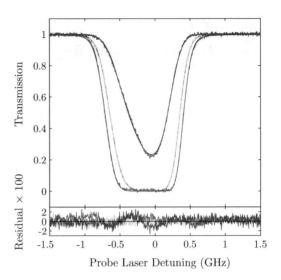

the vapour, and fluorescence leaving the system. In the second heater, the ceramic
resistors are placed directly in contact with the glass cell, and a Teflon case provides
insulation. The smaller heat capacity of the second design provides the advantage of
faster warming up.

Both cell heaters are regulated by a feedback loop to achieve a stable temperature.
The output of a thermocouple placed in contact with the quartz cell is read by a *Raspberry Pi®*, which then sets the power in the ceramic resistors. Using the *Raspberry Pi®* allows easy adjustment of the feedback parameters and the settings for each
cell heater may be saved and later recalled. Furthermore, non-linear feedback and
warm-up sequences can be implemented for time efficiency.

Although the thermocouple used for temperature stabilisation was placed in contact with the quartz glass cell, it did not accurately reflect the temperature of the
caesium vapour inside the cell. Instead, the vapour pressure was inferred by measuring the transmission line shape of the D_2 transition. Increasing the vapour pressure
of the cell leads to increased absorption of the laser beam, and hence broadening and
modification of the lineshape. By fitting the absorption lineshape, it is possible to
extract the temperature of the vapour. This is performed using a Python code, *Elecsus*, developed by other members of the Durham group [1], and in Fig. 3.1 we show
examples of the fitted line shapes. The discrepancy between the fitted temperature
and the thermocouple was typically around 2 °C.

3.2 Laser Systems

Figure 3.2 shows the energy levels and excitation scheme used for the experiments
in this thesis. The probe laser beam drives the caesium D_2 transition (852 nm), and is
generated with a *Toptica DL100* external cavity diode laser (ECDL). The coupling

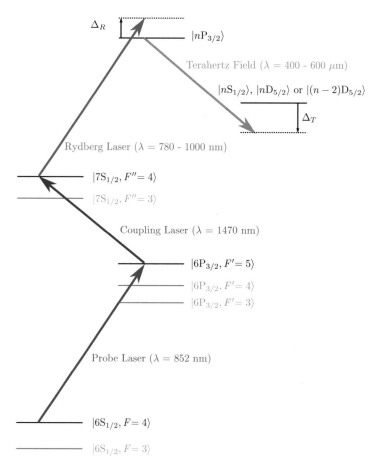

Fig. 3.2 Three step ladder excitation scheme. A probe, coupling and Rydberg laser drive a ladder excitation scheme to Rydberg states in caesium, and a terahertz-frequency field drives transitions between Rydberg states

beam is provided by an ECDL made by *Sacher Lasertechnik* and drives a transition from the $6P_{3/2}$ state to the $7S_{1/2}$ state (1470 nm). The Rydberg laser beam, generated by a *M-Squared SolsTis* Titanium Sapphire laser, drives atoms from the $7S_{1/2}$ state to nP Rydberg states, and the wide tunability of the laser system allows us to access Rydberg states from $n = 10$ to ionisation. We note that the Ti:Saph system could be replaced by a cheaper diode laser and tapered amplifier system, though at the expense of some tunability. Together, the lasers form a three-step ladder excitation scheme. A three-step scheme (rather than one or two laser beams) is an attractive method for exciting Rydberg states because the transition strength for each transition is larger, resulting in smaller power requirements. Furthermore, the infrared diode lasers are easier to use and maintain than the blue or ultra-violet lasers that would be necessary for two-step or one-step excitation. Finally, a Terahertz field is used to drive transitions between Rydberg states (see Sect. 3.2.3 for details on the THz source).

3.2.1   Laser Frequency Stabilisation

The probe laser beam is frequency stabilised using polarisation spectroscopy in a reference caesium cell [2, 3]. A circularly polarised pump beam induces frequency dependent birefringence, which is read out as a rotation of the polarisation of a weak, linearly polarised, counter-propagating beam. Using a differencing photodiode, a dispersive signal is generated with a zero crossing for each hyperfine transition, each with sub-doppler resolution. The $F = 4 \rightarrow F' = 5$ hyperfine transition is selected because it is both a closed transition, and it has the strongest transition strength. The signal is sent to the *Toptica* laser controller, which provides feedback to the position of the grating that forms the external cavity of the ECDL.

The coupling laser is frequency stabilised using excited state polarisation spectroscopy [4]. Frequency dependent birefringence at the coupling laser wavelength is generated by a circularly polarised D_2 pump beam. The birefringence is measured using a weak laser beam at the coupling wavelength, and another differencing photodiode is used to generate a dispersive signal. Feedback to the ECDL grating is controlled by a home-built circuit designed by Šibalić. Both polarisation spectroscopy and excited-state polarisation spectroscopy are conducted in a single 70 mm cell at room temperature, and a Mu-metal shield was used to exclude ambient magnetic fields.

3.2.2   Rydberg Laser

The *SolsTis* laser system was frequency stabilised to an internal reference etalon, which could be controlled by the user. However, when the etalon was scanned the laser frequency lagged behind the etalon, leading to hysteresis. In order to monitor the laser frequency, some of the laser light was sent to a second, external etalon, with free spectral range 300 MHz. As the Rydberg laser was scanned, the transmission through the external etalon was measured and the times of the transmission peaks were recorded. The relative laser frequency, f^{rel}, could then be deduced as a function of time, t, by fitting a polynomial function, $f^{\text{rel}} = \mathcal{P}(t)$, to the times of the transmission peaks. The order of the polynomial was selected according to the degree of nonlinearity of the scan and the number of transmission peaks that the scan included. A typical scan over 3 GHz would include 10 transmission peaks, of which the middle 8 were fitted with a third order polynomial. This method for removing hysteresis effects from the laser scan was essential when measuring the hysteresis response of the caesium vapour (Chaps. 5 and 8).

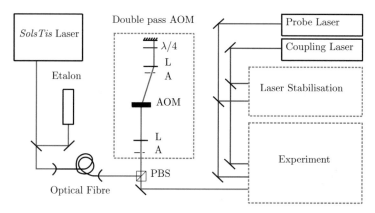

Fig. 3.3 Rydberg laser preparation. The Rydberg laser is derived from a *SolsTis* Ti:Sapphire laser, and passes through an optical fibre and double pass acousto-optic modulator set-up using a quater waveplate ($\lambda/4$), polarising beam splitting cube (PBS), apertures (A) and lenses (L). An etalon is used to monitor the frequency scan

Before arriving at the caesium vapour the Rydberg laser light went through both a polarisation-maintaining optical fiber and a double pass acousto-optic modulator (AOM). On each pass the AOM shifted the laser frequency by an amount set by the radio frequency (RF) driver, in the range 150–250 MHz. By adjusting the RF driver a fast frequency scan was achieved, although with limited scan range. The AOM was also used for fast intensity modulation, with a switching time $t^{\text{rise}} < 1\,\mu$s. A schematic of the Rydberg laser preparation is shown in Fig. 3.3. Both the internal etalon scan and the AOM control were automated using LabView, allowing repeated measurements in quick succession, and minimising the opportunity for system parameters to drift between measurements.

3.2.3 Terahertz Source

The terahertz-frequency field is generated by an Amplifier Multiplier Chain (AMC) manufactured by *Viginia Diodes Inc*, which uses a chain of diodes to multiply the frequency of a microwave signal by a factor of 54. The frequency of the microwave input ($\approx 9 \rightarrow 14$ GHz), and hence the frequency of the output ($500 \rightarrow 750$ GHz) is controlled through a LabView computer program. The power output of the THz beam is in the range $1 \rightarrow 15\,\mu$W, and depends on the frequency. The AMC has a TTL on/off control which may be operated up to 1 kHz, and a voltage-controlled internal attenuator which we calibrate in Chap. 6.

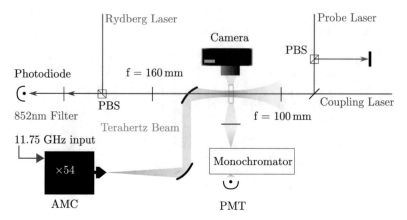

Fig. 3.4 Bench layout for Rydberg experiments: The probe, coupling and Rydberg lasers are combined using dichroic mirrors and polarising beam splitting (PBS) cubes, and focused through a 2 mm caesium vapour cell in a co-axial geometry. The vapour is analysed using a photodiode to monitor the transmission of the probe laser, and a camera and spectrometer to measure the atomic fluorescence. A terahertz beam is generated by an amplifier multiplier chain (AMC), and focused through the caesium vapour using gold-coated parabolic mirrors

3.2.4   Bench Layout

For the experiments described in Chaps. 5, 6, 7 and 8 all three laser beams were required to excite Rydberg atoms, which were subsequently manipulated with the terahertz beam. The layout used in these experiments is shown in Fig. 3.4. The three laser beams are combined using a sequence of polarising beam splitting (PBS) cubes and dichroic mirrors, and pass through the temperature stabilised caesium vapour cell. The diverging THz beam from the AMC is collimated and re-focused using a pair of gold-coated parabolic mirrors, with focal lengths 150 and 50 mm. When the Rydberg laser power exceeded 1 W, heating effects caused problems at locations where the beams were combined. It was found that the heating from the Rydberg laser changed the transmission or reflection of the light in the other laser beams. The problem was particularly severe with interference filters, and the PBS cubes were found to be the best available solution.

3.3   Experiment Read-Out

The dynamics of the caesium vapour were monitored by measuring both atomic fluorescence and probe laser transmission. The probe laser transmission is influenced by the Rydberg population either through EIT [5] or population shelving [6]. The latter mechanism causes an increase in probe laser transmission because atoms shelved in long-lived excited states are unable to absorb photons from the probe laser. We

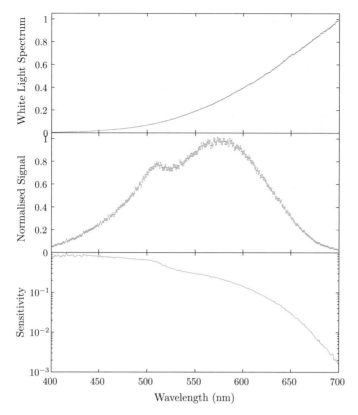

Fig. 3.5 Monochromator Calibration: We compare the known spectrum of a white light source (top) to a measurement of the same spectrum using our lab equipment (middle), and deduce the relative sensitivity of our equipment as a ratio between the two (bottom)

therefore interpret the increased laser transmission as an indication of the Rydberg population in the vapour. However the presence of any long lived state (or ionisation process) will have the same effect, and therefore increased probe laser transmission cannot have a strict correspondence to the Rydberg population. Nevertheless the transmitted probe beam power was measured using an avalanche photodiode (APD), and the fast, strong signal forms the principal read-out method of the vapour dynamics.

Atomic fluorescence at visible wavelengths (400–700 nm) is recorded by both a spectrometer and a camera. This range of wavelengths encompasses Rydberg state decay channels to the 5D states (red) and 6P states (green), and the blue decay from 7P to 6S (see Sect. 7.2.1). We use a *Canon* 5D Mk III camera with a macro lens to photograph the atomic fluorescence, gaining spatially-resolved information about the vapour. The RAW image files are interpreted using *Matlab* to ensure that the 14-bit CCD signal is not skewed by image processing software. The camera was set to the maximum sensitivity (*ISO* 25600), and the length scale of the images was calibrated by photographing a centimetre rule.

A monochromator and a photon-multiplier tube (PMT) were used to measure the spectrum of the visible atomic fluorescence. The wavelength transmitted through the monochromator was controlled with LabView, and the voltage from the PMT was passed to the LabView program using a analogue to digital converter (ADC). We derive a rough photon counting signal by recording the fraction of voltage samples that exceed a user-defined threshold. In order to calibrate the spectral sensitivity of the combined system, the spectrum of a white light was measured, and compared to a reference measurement made by a calibrated spectrometer (Fig. 3.5). Both the reference spectrometer and the monochromator-PMT system signals are proportional to photon count rate, γ, rather than the light intensity, I (which are related by $I = \gamma hc/\lambda$ where h is Planck's constant, c is the speed of light and λ is the wavelength of the fluorescence). The system is most sensitive in the range 400–600 nm, but falls off strongly above 600 nm, matching the specified sensitivity of the PMT.

3.4 Conclusion

We will go on to use the techniques outlined in this chapter to perform the experiments described in the rest of this thesis. In Chap. 4 we use the frequency-stabilised probe and coupling lasers together with a commercial photon counting unit to measure hyperfine quantum beats. The three-step ladder scheme is used in Chap. 5 to excite Rydberg atoms, which are then manipulated by a THz field in Chaps. 6, 7 and 8. The spatially resolved read out from the camera illuminates domain formation (Sect. 5.4) and spatial variation of a THz field (Chap. 7).

References

1. M.A. Zentile et al., ElecSus: a program to calculate the electric susceptibility of an atomic ensemble. Comput. Phys. Commun. **189**, 162 (2015)
2. C. Wieman, T.W. Hänsch, Doppler-free laser polarization spectroscopy. Phys. Rev. Lett. **36**, 1170 (1976)
3. C. Pearman et al., Polarization spectroscopy of a closed atomic transition: applications to laser frequency locking. J. Phys. B **35**, 5141 (2002)
4. C. Carr, C.S. Adams, K.J. Weatherill, Polarization spectroscopy of an excited state transition. Opt. Lett. **37**, 118 (2012)
5. C. Carr et al., Three-photon electromagnetically induced transparency using Rydberg states. Opt. Lett. **37**, 3858 (2012)
6. C. Carr, R. Ritter, C.G. Wade, C.S. Adams, K.J. Weatherill, Nonequilibrium phase transition in a dilute Rydberg ensemble. Phys. Rev. Lett. **111**, 113901 (2013)

Chapter 4
Probing an Excited State Transition Using Quantum Beats

We observe the dynamics of an excited-state transition in a room temperature atomic vapour using hyperfine quantum beats. The experiment consists of a pulsed excitation of the caesium D_2 transition, and continuous-wave driving of an excited-state transition from the $6P_{3/2}$ state to the $7S_{1/2}$ state. Quantum beats are observed in the fluorescence from the $6P_{3/2}$ state which are modified by the driving of the excited-state transition. The Fourier spectrum of the beat signal yields evidence of Autler-Townes splitting of the $6P_{3/2}$, $F = 5$ hyperfine level and Rabi oscillations on the excited-state transition. A detailed model provides qualitative agreement with the data, giving insight to the physical processes involved.

4.1 Introduction

Quantum beats are an elegant example of quantum interference, and their measurement has found an important application as a high resolution spectroscopy technique [1]. In an analogy to the two slit experiment where spatial interference is seen between two paths taken by a quantum particle, quantum beating is the temporal interference between photon scattering paths corresponding to different atomic states. In the way that the period of two-slit interference fringes is the reciprocal of the slit spacing, so the quantum beat frequency is related to the energy interval between atomic states. Therefore quantum beats reveal the *relative* energies of closely spaced atomic states. This is particularly useful for resolving spectral lines that otherwise overlap due to Doppler broadening. Applications have included the measurement of fine- [2] and hyperfine-[3] splittings, and Stark and Zeeman shifts [4, 5] in atoms and molecules [1, 4, 6].

This chapter describes work published:
'*Probing an excited-state atomic transition using hyperfine quantum-beat spectroscopy*',
C. G. Wade, N. Šibalić, J. Keaveney, C. S. Adams, and K. J. Weatherill,
Phys. Rev. A **90**, 033424 (2014).

In the process of developing a system to study the interaction of Rydberg atoms with terahertz electric fields, a pulsed ladder scheme was considered for exciting Rydberg states. Although the scheme was not adopted for the experiments described in this thesis (which rely on continuous wave driving), pulsed excitation became a topic of research in itself. During the development we observed hyperfine quantum beats caused by pulsed laser excitation of the D_2 line. Laser driving of the second step of our ladder scheme modifies the beat signal, allowing us to infer the dynamics of the excited-state transition. Excited-state transitions with large enough dipole transition moments can be probed directly [7], but more often electromagnetically induced transparency (EIT) is used. However, ladder EIT cannot be observed in our 'inverted wavelength' system where the upper transition wavelength (1470 nm) is longer than the lower (852 nm) [8, 9]. In this chapter we use quantum beats as an effective method to probe such 'inverted-wavelength' ladder systems in a thermal vapour.

4.1.1 Principle of Quantum Beats

We outline a toy model to illustrate the physical principle of quantum beats (Fig. 4.1, left column). The toy model consists of a zero-velocity atom with ground state $|g\rangle$, and two excited states $|e_1\rangle$ and $|e_2\rangle$ which are close in energy. If the transitions $|g\rangle \rightarrow |e_1\rangle$ and $|g\rangle \rightarrow |e_2\rangle$ are driven by a pulse with sufficient bandwidth to exceed the energy interval between the two excited states, a coherent superposition of the states is prepared. After the pulse each of the excited states accumulates phase according to its energy, evolving with the factor $e^{iE_i t/\hbar}$. The phase evolution produces a phase shift between the two superposed states, $\Delta\phi = \omega_b t$, causing the shape of the electron wavefunction to oscillate at the beat frequency, $\omega_b = (E_2 - E_1)/\hbar$. Because the coupling to vacuum modes is dictated by the shape of the electron wavefunction (see Sect. 2.2.3), the time-resolved fluorescence into an appropriately chosen mode, characterized by polarisation and propagation direction, is modulated by beating [1]. Nevertheless, the *total* decay rate remains constant, and the *total* fluorescence decays exponentially according to the excited state lifetime. These 'quantum beats' represent interference between the two different quantum pathways associated with $|e_1\rangle$ and $|e_2\rangle$. The interference is erased if information regarding which pathway was taken is recovered (e.g. spectroscopically resolving the fluorescence from each state).

The lifetime and energy state intervals of the caesium $6P_{3/2}$ hyperfine manifold make it a suitable system to probe using quantum beats. The interaction between the magnetic moment of the valence electron and the nuclear spin splits the $6P_{3/2}$ state into four energy levels with total angular momentum, $F = \{2, 3, 4, 5\}$ (Sect. 2.1.3). The energy interval between the $F = 5$ and $F = 4$ hyperfine states, $\Delta E_{5,4} = 2\pi\hbar \times 251$ MHz (Fig. 2.1), gives a beat period, $2\pi\hbar/\Delta E_{5\rightarrow4} = 4$ ns. The lifetime of the $6P_{3/2}$ state, $\tau = 30.57(7)$ ns [10], means that we can expect to measure 15 cycles of quantum beats if we observe the fluorescence over a duration of 2τ.

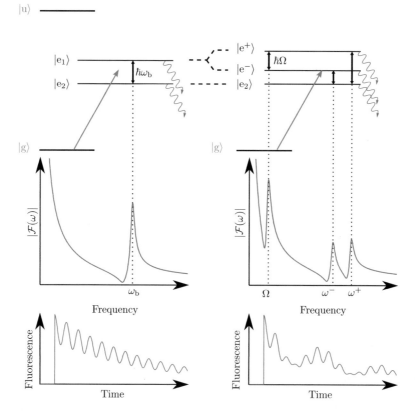

Fig. 4.1 Toy model. Left column: A short pulse prepares a superposition of states $|e_1\rangle$ and $|e_2\rangle$ which fluoresce to produce quantum beats. The Fourier transform (middle row) of the fluorescence signal (bottom row) allows the investigator to read off the beat frequency, ω_b. Right column: Driving an excited state transition is most easily understood in the dressed state picture. State $|e_1\rangle$ is coupled to state $|u\rangle$ to produce dressed states $|e^+\rangle$ and $|e^-\rangle$. The fluorescence signal now shows a trio of beat frequencies, Ω, ω^+ and ω^-

4.2 Caesium Hyperfine Quantum Beats

We describe an experiment to observe hyperfine quantum beats on the caesium D_2 line. The caesium vapour is maintained at room temperature (19 °C), and we excite the atoms using a probe laser stabilised to the D_2 transition (Sect. 3.2). The probe laser is modulated to create short pulses (1/e time ≈ 0.8 ns—Fig. 4.2), using a Pockels cell between high extinction polarisers. The pulses have a Fourier limited bandwidth which exceeds the energy seperation of the hyperfine states, and so the pulses prepare a coherent superposition of the $F = 3, 4, 5$ states which are dipole allowed from the $F = 4$ ground states. The fluorescence is measured using a single photon detector module which generates a TTL pulse for each photon. The pulses are timed

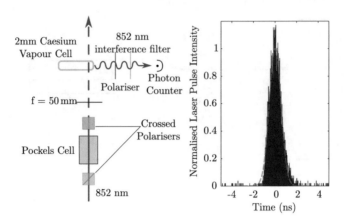

Fig. 4.2 Left: Time-resolved atomic fluorescence is recorded by a photon counting unit. Laser pulses are created by a Pockels cell which is placed between crossed polarisers. Right: Measurement of the pulse profile. The red dotted line is a Gaussian fit with 1/e time of 0.8 ns.

and counted by a high-bandwidth oscilloscope and in this way we achieve nanosecond timing resolution. To avoid saturating the counting module, we ensure that the expected delay between photons is much longer than the dead time of the counting module ($\approx$35 ns). In practice this means restricting the collection efficiency $\leqslant$1%. The caesium cell and photon counter were placed in a dark box to remove signal due to ambient light, and a 852 nm interference filter is placed in front of the detector. When the laser light arrives at the cell it has vertical polarisation, and we detect either horizontal or vertical polarised fluorescence propagating in the horizontal plane.

In the left column of Fig. 4.3 we present measurements of vertical and horizontal polarized fluorescence (top and bottom respectively). The modes show exponentially decaying fluorescence, modulated by beating. Consistent with our expectation that the total fluorescence is not modulated, the two modes beat out of phase. Initially the atomic dipole is aligned with the vertical laser polarisation, so we see that the vertical polarised fluorescence slightly precedes the horizontal fluorescence. In the right column of Fig. 4.3 we present the same data in frequency space. The top row shows the magnitudes of the normalized Fourier transforms of the two modes. We observe peaks at 251 and 452 MHz, corresponding to the $6P_{3/2}$ hyperfine splitting [11] (highlighted with vertical dashed lines). In the bottom row we subtract the two signals to attain a difference signal, removing frequency components relating to the exponential decay. However, because the beating of the two polarisation signals is out of phase we retain the quantum beat frequency components, allowing us to resolve a feature at 201 MHz, corresponding to the $F' = 3 \rightarrow F' = 4$ hyperfine splitting. The beating of the $F' = 3 \rightarrow F' = 4$ states is very weak compared to the other two hyperfine intervals because the population in these two states is limited. This restricted population is a result of

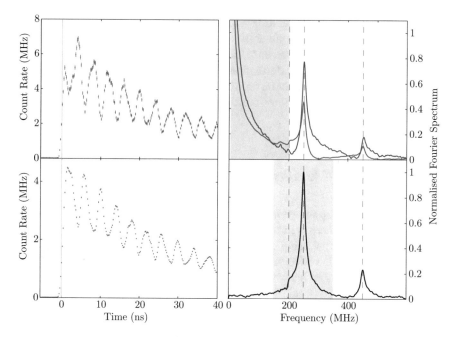

Fig. 4.3 Hyperfine quantum beats: Time resolved fluorescence measurements (left column) show the vertical (top) and horizontal (bottom) polarised fluorescence beating in antiphase (the center of the excitation pulse is incident at time $t = 0$ ns). The Fourier transform of the time signal shows the beat frequencies (right column). We show the magnitude of the Fourier transform (top) for the vertical (blue) and horizontal (red) polarised fluorescence and the difference signal (bottom)

both weaker coupling to the ground state and also detuning from the middle of the excitation pulse bandwidth which is centred on the $F = 4 \rightarrow F' = 5$ transition.

4.3 Quantum Beats Modified by Driving an Excited-State Transition

We describe a method to modify the caesium D_2 quantum beat signal by coupling the $6P_{3/2}$ state to the $7S_{1/2}$ state. To understand the effects of driving the excited-state transition we return to our toy model (Fig. 4.1, right column), to which we add a doubly-excited state, $|u\rangle$, and a coupling laser tuned to the transition from $|e_1\rangle$ to $|u\rangle$. The coupling laser has constant intensity throughout the experiment sequence: both during and after the pulse that drives the $|g\rangle \rightarrow |e\rangle$ transitions.

It is easiest to consider the dressed state picture, in which the coupling laser splits $|e_1\rangle$ into two dressed states, $|e^+\rangle$ and $|e^-\rangle$, separated by $\Delta E = \hbar\Omega$. The original beat at frequency ω_b, is split into two distinct beats with frequencies, $\omega^+ = \omega_b + \Omega/2$ and $\omega^- = \omega_b - \Omega/2$. Furthermore, a new beat frequency is introduced with frequency

Ω. This beat frequency relates to Rabi oscillations with atoms cycling on the excited-state transition. We note that unlike the initial quantum beat, this cycling leads to a modulation of the total fluorescence, not just a particular polarisation mode. The Fourier spectrum includes all information regarding Autler-Townes splitting [12] of the state $|e\rangle$ and Rabi oscillations on the excited-state transition, $|e\rangle \rightarrow |u\rangle$, allowing us to read out the excited-state transition dynamics, embedded in the radio-frequency (RF) quantum beat.

4.3.1   Experiment Results

The experimental set-up shown in Fig. 4.2 is repeated, but now an additional counter-propagating 1470 nm laser is introduced. The laser is locked to the $6P_{3/2}$ $F' = 5 \rightarrow 7S_{1/2}$ $F'' = 4$ transition using excited-state polarisation spectroscopy (Sect. 3.2 [13]). To best control the effects of driving the excited-state transition, it is desirable to minimize the spread of intensity of the coupling laser that the atoms experience. To achieve this, we only sample the centre of the coupling beam ($1/e^2$ radius 0.3 mm) where the intensity is most uniform, by virtue of tighter focusing of the preparation pulse ($1/e^2$ radius 0.06 mm).

We present the results of our measurements in Fig. 4.4. The top row shows the vertical and horizontal polarised fluorescence (left and right respectively), with the driving field intensity at the center of the laser beam $I_d = 4$ W cm^{-2}. The middle row shows the Fourier spectrum for the same data, with the two signals separately on the left and subtracted on the right. The bottom row shows colourplots covering a range of coupling laser intensities $I_d = 0 \rightarrow 7$ W cm^{-2}, constructed from nine individual sets of intensity measurements. On the left we show the modulus of the Fourier transform of the vertically polarized fluorescence measurements and on the right we show the modulus of the Fourier transform of the difference signal.

We can see the changes to the Fourier spectra that we expected from considering the toy model. First a new oscillation is present, leading to a peak in the Fourier spectrum at 100 MHz when the coupling laser has intensity $I_d = 4$ W cm^{-2}, and to the diagonal feature in the colour map of vertical polarised fluorescence spectra. The new oscillation represents atoms performing Rabi oscillations on the excited-state transition. However, the Rabi oscillation does not leave a footprint in the difference spectrum because it modulates the entire 852 nm fluorescence. Second, the peak in the spectrum relating to the $F' = 5 \rightarrow F' = 4$ beat (251 MHz) is split in a doublet, as seen clearly in the difference signal. The origin of this effect is Autler-Townes splitting of the $6P_{3/2}$ $F' = 5$ atomic state, caused by driving the excited-state transition. However, whilst the simple toy model predicts that the splitting of the beat frequency would be equal to the frequency of the Rabi oscillation, it is clear from the data that the splitting is smaller than the Rabi frequency. Furthermore, the higher frequency branch of the doublet is stronger than the low frequency branch. This effect is even more exaggerated in the $F' = 5 \rightarrow F' = 3$ (452 MHz) beat where we do not observe the low frequency branch at all. Both the absence of the lower branch and

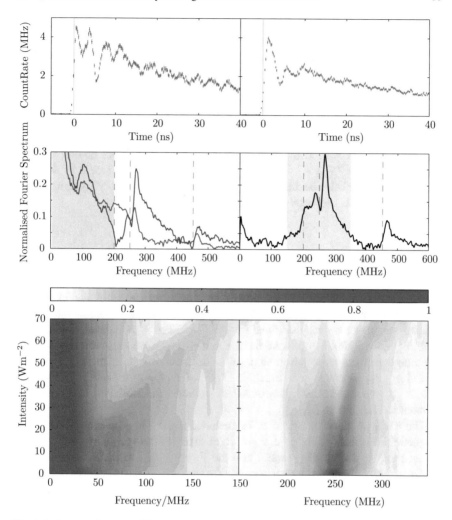

Fig. 4.4 Quantum beats modified by an excited state transition. Top row: For measured intensity $I_d = 4$ W cm^{-2} we show the vertical polarised (left) and horizontal polarised (right) fluorescence as a function of time. Middle row: The beat spectra are shown for vertical (blue) and horizontal (red) polarised fluorescence (left), and the difference signal (right). The dashed vertical lines show the bare hyperfine beat frequencies, and the shaded areas correspond to colourmaps (bottom row), showing the vertical polarised signal (left) and the difference signal (right) with changing coupling laser intensity.

the unexpected doublet splitting originate from Doppler shift effects that we explain using a comprehensive computer simulation in Sect. 4.3.2.

We note that the fluorescence decays more slowly as the longer lived $7S_{1/2} F'' = 4$ state is mixed into the $6P_{3/2}$ states, and the total amount of measured 852 nm fluorescence decreases. The slower decay is consistent with literature values of the $7S_{1/2}$ lifetime (49 ± 4 ns [14]). However, because the atoms are periodically switching between the $7S_{1/2}$ and $6P_{3/2}$ states it is not possible to use a simple rate equation to model the decay. Instead it is necessary to compare the data to the full computer simulation (Sect. 4.3.2). The decrease in total fluorescence might be due to atoms decaying from the $7S_{1/2} F'' = 4$ state via the $6P_{1/2}$ manifold instead of the $6P_{3/2}$ manifold, but could also be due to hyperfine optical pumping caused by light leaking through the Pockels cell between pulses.

4.3.2 Computer Simulation

For a more rigorous analysis of the experiment results we use a computer simulation to model the atomic fluorescence. The simulation is based on the formalism for treating spontaneous decay outlined in Sect. 2.2.3, and is calculated in two steps: First the optical Bloch equations for the system are solved numerically; second, the time-dependent expectation value of a 'detection operator' relating to the D_2 transition, $\mathcal{L}_{e_d}$, is calculated (Eq. 2.27). The calculation process is repeated for a sample of velocity classes which are then summed and weighted according to a Boltzmann distribution.

The basis of states that need to be included is sizable—we require all of the magnetic sub-levels (m_F) of several groups of states: the two $6S_{1/2}$ hyperfine ground states; three hyperfine $6P_{3/2}$ states and the F = 4 $7S_{1/2}$ state: 52 magnetic sub-levels in total. Solving the master equation for such a large basis requires prohibitive computing capacity, so some approximations were made. We note that the lasers are linearly polarised with parallel electric fields and so if we choose a quantisation axis along the field direction the lasers only drive π transitions. Whilst this basis might not be the energy eigenbasis due to uncompensated laboratory magnetic fields, any dynamics as a consequence of this can be neglected since the duration of our experiment is much shorter than the relevant Larmor precession timescale. Therefore m_F is conserved for each atom during the excitation process, and so we divide the full 52 state problem into nine m_F subspaces, each including one state from each of the hyperfine levels: $6S_{1/2} F = 3, 4$; $6P_{3/2} F' = 3, 4, 5$ and $7S_{1/2} F'' = 4$.

The Rabi frequencies for each subspace are calculated individually by decomposing the reduced dipole matrix according to the formalism described in Sect. 2.2.1. We evolve the density matrix, $\hat{\rho}_{m_F}$, representing each of the nine m_F subspaces, according to the master equation (Eq. 2.21), saving the density matrix at each time step. In the final process of our model, we collate the populations and coherences from the nine subspaces into a single density matrix for each time step, giving the complete state of the atom as it changes in time. Using the 'detection operator', we project the atomic dipole at each time step and hence infer both the linearly and circularly polarized 852 nm fluorescence. The convenient sub-division offers a computational speed up that permits the simulation to be run on a desktop computer.

Although the subspaces are not coupled by the driving laser fields, we note that they are not truly separate since atoms can undergo spontaneous $\sigma^{\pm}$ transitions resulting in a change of m_F quantum number. Instead of modeling the full behavior, we attribute the total rate of spontaneous decay of each state to π transitions when solving the master equation, thus conserving the total population in each subspace. Whilst this approximation leads to a small error in the solution of the master equation, it does not prevent us from calculating the full quantum beat signal from the collated atomic state (including $\sigma^{\pm}$ fluorescence), as can be seen from the following justification.

Decay via $\sigma^{\pm}$ transitions can occur on either the $6P_{3/2} \rightarrow 6S_{1/2}$ or the $7S_{1/2} \rightarrow 6P_{3/2}$ transition. In the first case, consigning the full decay rate to π-transitions leads to an error in the calculated population of the ground state hyperfine levels. However, the rate of $\sigma^{\pm}$ fluorescence is still correctly inferred by the 'detection operator' because it is only sensitive to the populations and coherences of the $6P_{3/2}$ manifold. As these only depend upon the *total* decay rate, the correct fluorescence signal is recovered by the 'detection operator'.

In the second case the same approximation leads to a small error in the populations in the $6P_{3/2}$ manifold. However, because atoms decaying into this level do not contribute to the atomic *coherences*, the modulation of the D_2 fluorescence (eg. the beating) is still calculated correctly, and we would only expect the approximation to contribute a small error in the incoherent background (exponentially decaying fluorescence), caused by unbalanced exchange of population between neighboring m_F subspaces.

4.3.3 Analysis

In this section we make a direct comparison between the computer simulation and the measured data. In Fig. 4.5 we present the results of the vertical polarized fluorescence for both the experiment and simulation. The unperturbed hyperfine quantum beat signal fits well and, although the features are often more pronounced in the simulation than the data, we see at least qualitative agreement for the perturbed beats. On the strength of this we can draw additional physical insight about the system.

In Sect. 4.3.1 we noted that the splitting of the $F' = 5 \rightarrow F' = 4$ hyperfine quantum beat was unexpectedly smaller than the measured frequency of the Rabi oscillation. We suggest this is similar to narrowed EIT windows in thermal vapours [16, 17] where off-resonant velocity classes partially fill the transparency window left by resonant atoms. Calculated contributions to the splitting of the $F' = 4 \rightarrow F' = 5$ quantum beat from different velocity classes are shown in Fig. 4.6. The zero-velocity class (bold, green) shows a splitting that is consistent with the simulated excited-state transition Rabi frequency, yet this is much larger than the splitting which appears in the total signal. However, contributions from off-resonant velocity classes fill in

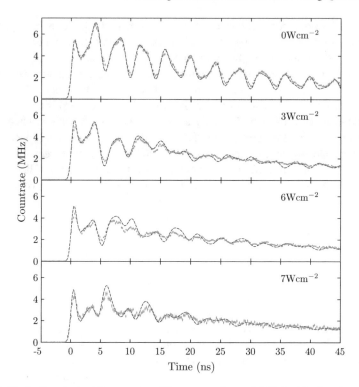

Fig. 4.5 Vertically polarized fluorescence: We compare the model (dashed line) and experimental data points for measured excited-state transition laser intensity $I_d = (0, 3, 6, 7)$ W cm^{-2} (top to bottom). We note that the visibility of the peaks is always smaller than the model predicts. The incident light pulse occurs at time $t = 0$ ns and the error bars are calculated from Poissonian photon counting statistics [15]

the gap. The inset compares this best-fit calculated spectrum with our data and we see qualitative agreement, although we acknowledge a significant discrepancy in the simulated ($I_d^{Sim} = 0.9$ W cm^{-2}) and measured ($I_d = 3$ W cm^{-2}) excited-state transition laser intensities.

We also noted in Sect. 4.3.1 that the high-frequency branch of the split $F' = 4 \rightarrow F' = 5$ quantum beat makes a stronger contribution to the Fourier spectrum than the low-frequency branch. This unexpected asymmetry can be explained by constructing a two-step argument: First, we note that the strongest beats arise from atoms experiencing a red Doppler shift of the 852 nm laser. This moves the center of the frequency profile of the pulse between the two beating transitions, promoting the excitation of both levels as required for quantum beats. Second, the atoms which experience a red-shift for the 852 nm laser see a blue-shift of the 1469 nm laser because the laser beams are counter-propagating. This blue-shift means that the dressed states represented in the higher frequency branch of the split quantum beat have a greater admixture of the 6P$_{3/2}$ $F' = 5$ state, and as such a stronger coupling to

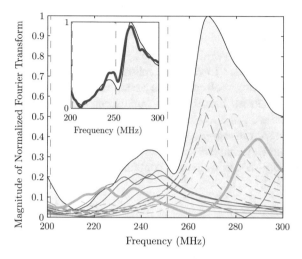

Fig. 4.6 Calculated magnitude of the normalized Fourier transform of the vertically polarized fluorescence: We show a break down of velocity class contributions, with dashed (red) lines showing individual velocity classes travelling away from the 852 nm laser, and solid (blue) lines showing velocity classes traveling towards the 852 nm laser. The velocity classes are spaced at $33\,\mathrm{ms^{-1}}$ intervals and the bold (green) line shows the contribution from zero-velocity atoms. The gray shaded area shows the scaled sum of the signals and the calculation relates to probing a region with a uniform excited-state transition driving field intensity $I_{\mathrm{d}}^{\mathrm{Sim}} = 0.9\,\mathrm{W\,cm^{-2}}$. The inset compares the data taken for measured CW driving field intensity $I_{\mathrm{d}} = 3\,\mathrm{W\,cm^{-2}}$ shown in bold (blue) and the summed model (gray shaded area and black line)

the ground state. Thus the imbalance between branches of the quantum beat comes from a bias towards a particular velocity class, and a bias within this velocity class to a particular branch. The asymmetry between branches is even stronger for the $F' = 3 \rightarrow F' = 5$ quantum beat as the $F' = 3$ hyperfine state is further in energy from the $F' = 5$ hyperfine state. Consequently, only the high frequency branch of the splitting was observed and the lower-frequency branch is absent (Sect. 4.3.1).

There are some remaining discrepancies between the simulation and the data. We found that when we used the CW driving field intensity as a fit parameter in the model, the best fit did not match or even scale linearly with the intensity we measured in the experiment. We believe that this might originate from optical pumping between excitation pulses. The high extinction polarizers each side of the Pockels cell (Fig. 4.2a) still allowed a few hundred nanowatts of 852 nm light to leak into the vapour cell for the 1 ms duration between pulses. For resonant velocity classes, this could have lead to an initial state other than the uniform distribution over the ground states that the computer simulation assumes.

4.4   Conclusion

We have demonstrated a novel method using hyperfine quantum beat spectroscopy to observe sub-Doppler Autler-Townes type splitting in an 'inverted wavelength' ladder scheme which would not be observable in a continuously excited room-temperature vapour. A comprehensive model of the fluorescence gives qualitative agreement with our data, and we use it to gain physical insight into the process. Our method offers a new means for investigating excited-state transitions in a room-temperature vapour. In the following chapters we extend the ladder system to a 3-photon Rydberg excitation scheme, although we pursue continuous wave (CW) excitation.

References

1. S. Haroche, *Quantum Beats and Time-Resolved Fluorescence Spectroscopy*, vol. 13 of *Topics in Applied Physics*, Chap. 7 (Springer-Verlag, 1976), p. 253
2. S. Haroche, M. Gross, M.P. Silverman, Observation of fine-structure quantum beats following stepwise excitation in sodium *D* states. Phys. Rev. Lett. **33**, 1063 (1974)
3. S. Haroche, J. Paisner, A. Schawlow, Hyperfine quantum beats observed in Cs vapor under pulsed dye laser excitation. Phys. Rev. Lett. **30**, 948 (1973)
4. E. Hack, J.R. Huber, Quantum beat spectroscopy of molecules. Int. Rev. Phys. Chem. **10**, 287 (1991)
5. J.N. Dodd, R.D. Kaul, D.M. Warrington, The modulation of resonance fluorescence excited by pulsed light. Proc. Phys. Soc. **84**, 176 (1964)
6. R.T. Carter, J.R. Huber, Quantum beat spectroscopy in chemistry. Chem. Soc. Rev. **29**, 305 (2000)
7. M. Tanasittikosol, C. Carr, C.S. Adams, K.J. Weatherill, Subnatural linewidths in two-photon excited-state spectroscopy. Phys. Rev. A **85**, 033830 (2012)
8. J. Boon, E. Zekou, D. McGloin, M. Dunn, Comparison of wavelength dependence in cascade-, Λ-, and Vee-type schemes for electromagnetically induced transparency. Phys. Rev. A **59**, 4675 (1999)
9. A. Urvoy et al., Optical coherences and wavelength mismatch in ladder systems. J. Phys. B **46**, 245001 (2013)
10. R.J. Rafac, C.E. Tanner, A.E. Livingston, H.G. Berry, Fast-beam laser lifetime measurements of the cesium $6p^2 P_{1/2,3/2}$ states. Phys. Rev. A **60**, 3648 (1999)
11. D. Das, V. Natarajan, *Hyperfine spectroscopy on the* $6P_{3/2}$ *state of 133 Cs using coherent control*. Europhys. Lett. (EPL) **72**, 740 (2005)
12. S.H. Autler, C.H. Townes, Stark effect in rapidly varying fields. Phys. Rev. **100**, 703 (1955)
13. C. Carr, C.S. Adams, K.J. Weatherill, Polarization spectroscopy of an excited state transition. Opt Lett **37**, 118 (2012)
14. J. Marek, Study of the time-resolved fluorescence of the Cs–Xe molecular bands. J. Phys. B Atomic Mol. Phys. **10**, L325 (1977)
15. I. Hughes, T. Hase, *Measurements and Their Uncertainties* (Oxford University Press, 2010)
16. S. Iftiquar, G. Karve, V. Natarajan, Subnatural linewidth for probe absorption in an electromagnetically-induced-transparency medium due to Doppler averaging. Phys. Rev. A **77**, 063807 (2008)
17. M.G. Bason, A.K. Mohapatra, K.J. Weatherill, C.S. Adams, *Narrow Absorptive Resonances in a Four-level Atomic System*. J. Phys. B **42**, 075503 (2009)

Chapter 5
Intrinsic Rydberg Optical Bistability

Alkali metal atomic vapours excited to high lying Rydberg states exhibit intrinsic optical bistability which has been the subject of several recent studies [1–4]. In this chapter we explore phenomena associated with Rydberg optical bistability, and compare the experimental results with simple phenomenological models. We find that critical slowing down is absent around the critical point corresponding to the onset of bistability. The presence of a spatial phase boundary is observed, which we describe with a 1D interacting-chain model. The work lays preparation for Chap. 8, in which we study Rydberg optical bistability modified by a THz field.

5.1 Introduction

Bistability refers to a phenomenon whereby a system has two, distinct and stable responses to the same stimulus. A simple example of a system showing a bistable response is the Duffing oscillator [5], where anharmonicity in the restoring force of a mechanical system leads to an amplitude-dependent resonant frequency. When the oscillator is driven at a suitable off-resonant frequency, the system can exhibit either of two responses at the driving frequency: Either a low amplitude response caused by the mis-match between the resonance and the driving frequency, or a large amplitude response where the amplitude-dependent resonant frequency is shifted towards the driving frequency. In the optical domain, bistability can also be observed in simple systems such as the interface between a linear and a Kerr medium [6]. The refractive index of the Kerr medium depends on the intensity of the transmitted light,

This chapter includes work published:
'*Intrinsic optical bistability in a strongly driven Rydberg ensemble*',
N. R. de Melo, C. G. Wade, N. Šibalić, J. M. Kondo, C. S. Adams, and K. J. Weatherill,
Phys. Rev. A **93**, 063863 (2016)

which itself depends back upon the refractive index, resulting in a bistable response. This positive feedback cycle is characteristic of bistable systems, which typically require both non-linearity and feedback. Optical bistability has been observed in nematic liquid crystals [7], nonlinear prisms [8], photonic crystal cavities [9] and QED cavities [10], amongst other systems.

Although atoms show a non-linear, resonant response to optical driving fields, isolated single atoms cannot constitute bistable systems because quantum fluctuations preclude steady-state behavior [11, 12]. Instead small quantum systems must be described through quantum-jump or other probabilistic methods [13]. The response of a bulk atomic vapour can only exhibit bistability if there is a feedback mechanism, causing collective behaviour amongst its constituent atoms. This feedback might be provided by a Fabry-Perot cavity [14], or by interactions between the atoms [15, 16]. The latter phenomenon is known as intrinsic optical bistability, and it has been realised in Rydberg vapours [2], and Yb^{3+} ions embedded in a crystal at cryogenic temperatures [17].

5.1.1   Rydberg Level Shift

The feedback responsible for intrinsic Rydberg optical bistability is a population dependent energy shift of the Rydberg level. Analogous to the Duffing oscillator, Rydberg bistability is observed when the energy level shift brings the resonance towards the laser driving frequency. However, the microscopic origin of the level shift is not clear.

A recent study performed by Weller et al. advocates that the presence of ions might be responsible for the level shift [1]. It is known that ionisation occurs in high density room temperature Rydberg ensembles [18]. The hypothesis suggests that Rydberg atoms are ionised, causing electric fields within the vapour. In turn the electric fields polarise the Rydberg atoms, leading to a level shift. Combining measurements of different Rydberg states, Weller found that the sign of the level shift follows the sign of the Rydberg state polarisability. Furthermore, Weller undertook an experiment exciting two different Rydberg states at once in the same region of the vapour. Strong driving of the lower level led to broadening and shifting of an EIT signal measured with the higher level. This is only consistent with ions being present, as the disparity between the two energy levels is expected to completely suppress other mechanisms for the Rydberg level shift such as dipole-dipole interactions.

Measuring atomic fluorescence can also provide insight to the processes occurring in the vapour. In weak contradiction to the proposal that ions are responsible for the level shift, the fluorescence spectrum shows very little content from Rydberg states towards the ionisation energy [2]. This might be an indication that these high-lying Rydberg states are absent in the vapour, and that therefore so are the processes of ionisation and ion-recombination. Nevertheless, it is not clear that such high-lying Rydberg states would be sure to leave a signature in the fluorescence spectrum, particularly if they ionise much more quickly than undergoing optical decay.

Evidence for a superradiant cascade has also been observed in the fluorescence spectrum. Superradiance is a cooperative phenomenon which promotes decay via wavelengths longer than the inter-atomic spacing [19]. Thus dense ensembles of Rydberg atoms collectively cascade through closely spaced energy levels instead of decaying individually at optical wavelengths [20]. Exciting high densities of Rydberg atoms in thermal vapour cells has been shown to result in a suppression of fluorescence from the state that is being excited. Instead fluorescence from lower lying Rydberg states is observed, consistent with a superradiant cascade [2, 21]. A regime in which superradiance is present indicates that dipole-dipole interactions between the Rydberg atoms might play a role in the level shift.

5.1.2 Chapter Outline

In this chapter we explore Rydberg optical bistability using both experiment and theory. We begin with an example, outlining the techniques used for an experimental demonstration of optical bistability (Sect. 5.2). In Sect. 5.3 we measure a phase diagram and the system response time, which we compare to phenomenological models. Finally, we report the observation of a phase boundary, and extend the model to account for this behavior (Sect. 5.4).

5.2 Optical Bistability Example

In this section we present an example of optical bistability. The techniques described in Chap. 3 are used to excite Rydberg atoms in a 2 mm caesium vapour cell. A probe laser (852 nm) and coupling laser (1470 nm) are locked to their respective atomic transitions ($6S_{1/2} \rightarrow 6P_{3/2}$ and $6P_{3/2} \rightarrow 7S_{1/2}$), and together with a Rydberg laser (882 nm) excite atoms to the $12P_{3/2}$ Rydberg state. The probe, coupling and Rydberg laser beams have 3 mW, 4.5 mW and 382 mW power and beam waists 30 μm, 80 μm and 80 μm respectively, and the caesium vapour is maintained at 95 °C using a heating oven. We scan the detuning of the Rydberg laser and measure both the probe laser transmission and atomic flourescence (Fig. 5.1). The transmission signal shows increased transparency at the atomic resonance due to population shelving [2], and so we interpret the transmission as an indication of the Rydberg population. The form of the signal is similar to those reported in previous work [1, 2], and we see a single, asymmetric peak with a hysteresis cycle located on the side of negative detuning. At each side of the hysteresis loop an abrupt change occurs, which we identify as the Rydberg phase transition [2].

We measure the atomic fluorescence originating from the $12S_{1/2} \rightarrow 6P_{1/2}$ and $11D_{5/2}, 11D_{3/2} \rightarrow 6P_{3/2}$ transitions using a monochromator and photon multiplier tube (PMT). However, the strong fluorescence intensity precludes photon counting (Sect. 3.3), and instead we average the voltage output of the PMT over several scans

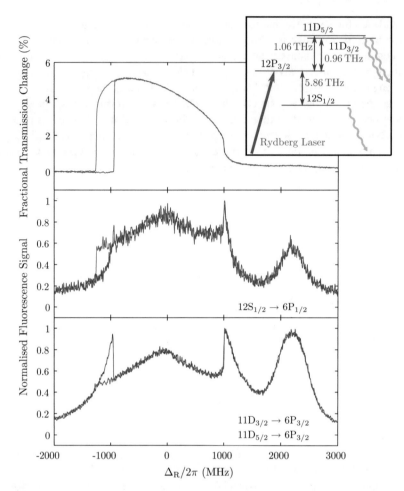

Fig. 5.1 Transmission and fluorescence signals as the Rydberg laser is scanned across the $7S_{1/2}$ to $12P_{3/2}$ transition (882 nm). The laser is scanned with increasing (blue) and decreasing (red) frequency. We show the measured atomic fluorescence on the $12S_{1/2} \rightarrow 6P_{1/2}$ (middle, 541 nm) and $11D \rightarrow 6P_{3/2}$ (bottom, 550 nm) transitions. We note that the spectrometer cannot resolve the 11D fine structure

using an oscilloscope. The 11D and 12S states are the closest in energy to the $12P_{3/2}$ state excited by the laser, lying above and below respectively. The flourescence signals have a very different form to the transmission signal, and we note some extra features. We see an almost triangular peak at ≈ 1 GHz, and a rounded peak centered ≈ 2.2 GHz. The frequency of the rounded peak matches the hyperfine splitting of the $7S_{1/2}$ state, and so we attribute this feature to atoms excited from the $7S_{1/2}$, $F = 3$ state by the Rydberg laser, but the origin of the triangular peak is unclear.

In the two branches of the hysteresis loop the fluorescence is very different, and the two emission lines we consider here exhibit opposite behaviour. The $12S_{1/2} \rightarrow 6P_{1/2}$

fluorescence is stronger when the probe laser transmission is greater, yet the $11D_{5/2}$, $11D_{3/2} \rightarrow 6P_{3/2}$ fluorescence is *weaker*. More generally the fluorescence appears bright orange to the eye when there is a high Rydberg density, but faint green for a low Rydberg density, and both colours include a large collection of transition lines [2]. The origin of the complicated spectrum is population re-distribution in the Rydberg manifold. Mechanisms which might play a role include collisions, superradiance, spontaneous decay and transitions induced by black-body radiation. Further study of emission lines might elucidate the dynamics of the vapour, and inform the debate concerning which mechanisms dominate the population re-distribution and the Rydberg level shift. We use photographs of the contrasting orange and green fluorescence to infer the spatially resolved Rydberg number density in Sect. 5.4.

Using the Ti:Saph laser it is possible to excite Rydberg states with principal quantum number from 10 up to the ionisation threshold, and so the example presented here is close to one end of the parameter space. As the principal quantum number increases, smaller number density is required to observe optical bistability, but the reduced number density results in weaker signals. Most of the work in this thesis is conducted using the $21P_{3/2}$ state, partly because the $21P_{3/2} \rightarrow 21S_{1/2}$ transition falls within the frequency range of the THz source, but also because it provides a compromise between signal strength and Rydberg interaction strength.

5.3 Phenomenological Modeling

As the discussion so far shows, the caesium Rydberg vapour has too much complexity for a complete model. It is therefore necessary to make approximations, a task which has been undertaken in several recent studies. Šibalić et al. used Monte-Carlo methods to simulate a sample of interacting 2-level atoms as they move past each other in the vapour [22]. The method does not require any mean-field approximation, although it is computationally intensive. The work investigates various interaction potentials, and finds that the motion of the atoms can act to stabilise the collective states.

In contrast, Marcuzzi et al. use a mean-field approximation to describe an ensemble of motionless, 2-level atoms [23]. Although this model is very simple, it has some important merits: the model not only reproduces the main characteristic features of the system, including optical bistability and a first order phase transition, but the steady-state population in the excited state can be expressed as the solution to a single cubic equation, making the model fast to solve and easy to work with. However the model assumes the validity of a mean field which, as Šibalić shows, is only suitable for particular classes of interactions.

In this section we compare the model used by Marcuzzi with experimental observations. The model comprises an ensemble of atoms with ground state $|g\rangle$ and excited (Rydberg) state $|e\rangle$. The ensemble is described by a density matrix which follows the master equation (Eq. 2.21). The Rydberg level shift is included in the Hamiltonian (Eq. 2.20) by writing $\Delta \rightarrow \Delta - V\rho_{ee}$, where ρ_{ee} is the population of the Rydberg state, and V is an interaction parameter. Using Eq. 2.21, the resulting non-linear optical Bloch equations for a 2-level atom are written,

$$\dot{\rho}_{ge} = i\Omega(\rho_{ee} - 1/2) + i(\Delta - V\rho_{rr})\rho_{ge} - \frac{\Gamma}{2}\rho_{ge}, \tag{5.1a}$$

$$\dot{\rho}_{ee} = -\Omega\text{Im}(\rho_{ge}) - \Gamma\rho_{ee}, \tag{5.1b}$$

where V is the interaction parameter, Ω is the laser Rabi frequency, Δ is the laser detuning, Γ is a phenomenological decay rate, and ρ_{ge} is the off-diagonal element of the density matrix ρ. We set $\dot{\rho}_{ge} = \dot{\rho}_{ge} = 0$ for the steady state, and eliminate ρ_{ge} to write a cubic equation in ρ_{ee},

$$\frac{\Omega^2}{4} - \left(\frac{\Omega^2}{2} + \frac{\Gamma^2}{4} + \Delta^2\right)\rho_{ee} - 2\Delta\rho_{ee}^2 - V^2\rho_{ee}^3 = 0, \tag{5.2}$$

which is solved numerically.

5.3.1 Steady State Phase Map

For a given interaction parameter, V, the mean-field model predicts a phase map in Ω and Δ, which we compare with experimental results in Fig. 5.2. As discussed in Sect. 3.3, we use the probe laser transmission as an indication of the Rydberg population, and so we compare ρ_{rr} with the fractional change in laser transmission. Where the system is monostable the phase diagram is shaded red; where the system is bistable the phase diagram is shaded blue. We note some simple similarities: in both the theory and experiment, a bistable region is present for negative detuning, over a range of laser powers. We will use the coordinates (Δ^c, Ω^c) to refer to the critical point in the phase diagram where bistability is present for the minimum possible laser power. However, the model also has a maximum Rabi frequency for which bistability can be present, a result that has not yet been observed in the lab. Possibly this is because of limited available laser power, but it could also indicate a discrepancy between the model and the system it describes. In general the phase map depends on many experimental parameters, including the power of the probe and coupling lasers along with the temperature of the cell, and the Rydberg state being excited. For the example presented in Fig. 5.2, the probe and coupling laser beams had $1/e^2$ radii and incident power $\{65\,\mu\text{m}, 35\,\mu\text{W}\}$ and $\{50\,\mu\text{m}, 44\,\mu\text{W}\}$ respectively, and the cell temperature was $79\,°\text{C}$.

A systematic study of the critical point with changing Rydberg state was made, and we found that the critical detuning follows the form $\Delta_R^c \propto (n - \delta)^\alpha$ where n is the principal quantum number, δ is the quantum defect and we measure $\alpha = 3.8 \pm 0.3$ (Fig. 5.3) [3]. Using this result, phenomenological formulas for Γ and V,

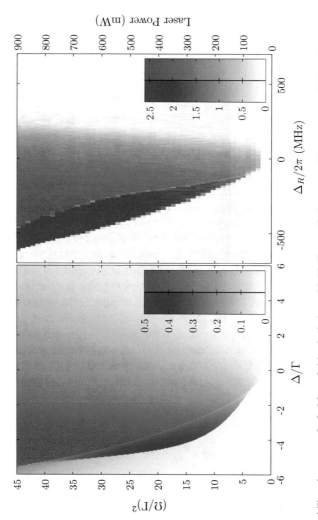

Fig. 5.2 Rydberg bistability phase map. Left: Mean-field calculation with $V/\Gamma = -10$. In monostable regions (red) the colourmap shows the excited state population, ρ_{ee}. In bistable regions (blue) the colourmap shows the difference between the two stable states, $\rho_{ee}^{high} - \rho_{ee}^{low}$. Right: Experiment result using state $28P_{3/2}$. In monostable regions (red) the colourmap shows the fractional transmission change in the probe beam (%). In bistable regions (blue) we show the difference signal. The Rydberg laser had a beam waist 60 μm giving a maximum Rabi frequency $\Omega_{max} = 2\pi \times 0.1\,\text{GHz}$

Fig. 5.3 Shift of the critical detuning, Δ_R^c, with changing principal quantum number, n. The data were recorded and analysed by N. R. de Melo

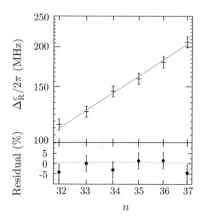

$$V \rightarrow V_0 n^{*4} N, \tag{5.3a}$$

$$\Gamma \rightarrow \Gamma + \beta n^{*4} N, \tag{5.3b}$$

where N is the density of the ground state atoms, $n^* = n - \delta$ and V_0 and β are coefficients, were used to compare the mean-field model with a range of Rydberg states, laser power and cell temperature. For each set of parameters the laser was scanned, and the range of detuning for which bistability is present ('bistability width') was recorded. It was found that as the principal quantum number of the Rydberg state increased, the bistability width also increased before going through a maximum.

The scaling of V and Γ with the 4th power of n^* hints at the potential mechanism causing the bistability. However, there are several phenomena that share n^{*4} scaling, including the ionisation cross section for collisions between pairs of Rydberg atoms [24], and the collisional quenching cross section between ground state atoms and Rydberg atoms [25]. If the presence of ions is invoked to explain the energy level shift an extra term going as the Rydberg polarisability (n^{*7}) should also be accounted for.

5.3.2 Critical Slowing Down

At the edge of the bistable parameter space, the system undergoes a non-equilibrium phase transition which is manifest as an abrupt change in the system properties. When the system changes phase it exhibits a delayed response known as critical slowing down. The delay time diverges algebraically near the phase transition, $\tau \propto (\alpha - \alpha^0)^{-\theta}$, where α is a general control parameter with value α^0 at the phase transition. The critical exponent calculated by Marcuzzi, $\theta = 1/2$ [23], matches measurements by Carr [2]. However, Marcuzzi goes on to predict a modification of the critical exponent at the critical point (Δ^c, Ω^c), changing from $\frac{1}{2}$ to $\frac{2}{3}$. In this section we perform an experiment to test this prediction.

The vapour is initialised with the probe and coupling lasers on, before the Rydberg laser is switched on with rise time $t_{swtich} < 1\,\mu s$, and we record the time-resolved probe laser transmission signal which is fitted to a hyperbolic tangent function,

$$s = s_0 + A \left(1 + \tanh\left((t - \tau)/T\right)\right), \tag{5.4}$$

where τ is the fitted delay time, T is the switching rate and A is the step size of the photodiode signal measuring the probe laser transmission. We repeat the experiment for different Rydberg laser power and detuning, and group the data according to the laser detuning. Each group covers a range of final Rydberg Rabi frequencies $\Omega_R < \Omega_R^0 \rightarrow \Omega_R > \Omega_R^0$, where Ω_R^0 is the Rabi frequency necessary to drive the phase transition. In general, Ω_R^0 depends on Δ_R. In Fig. 5.4 we present four groups of results, $(\Delta_R - \Delta_R^c)/2\pi = \{-90, -20, 0, 10\}\,MHz$. For each group we show the fitted delay time, τ, and the step size, A, as a function of the laser power (top and middle rows respectively). In the bottom row we show a selection of the time-resolved photodiode signals for $\Omega_R < \Omega_R^0$ (green) and $\Omega_R > \Omega_R^0$ (blue), with the fit functions, $s(t)$.

When $\Delta_R \ll \Delta_R^c$ (left column), we see the response time diverging at the critical power, and the step size, A, also shows a discontinuous response. The diverging response time is similar to the critical slowing down observed in a previous studies conducted by Carr et al. [2]. However, when $(\Delta_R - \Delta_R^c)/2\pi = -20\,MHz$, the response time does not diverge, and we only see a small increase of the fitted delay time, τ, in the data. This result is inconsistent with the prediction of Marcuzzi, which anticipates a diverging response time with critical exponent $\theta = \frac{2}{3}$ [23], and because our data does not show divergence we do not fit a value of θ. The reason for the discrepancy might originate from the detail of the microscopic dynamics. However, we note a crucial difference between this data and the results of Carr who demonstrated a clear step in the photodiode signal, with $T \ll \tau$ (see Fig. 4 in [2]). In contrast, the data presented here show a response which is not just delayed, but also slower, $T \approx \tau$. In order to reproduce the dynamics observed by Carr the author found it necessary to set the laser frequency far away from the critical detuning.

5.4 Phase Boundary

Photographing the atomic fluorescence can give an indication of the Rydberg number density along the laser beam. As we noted in Sect. 5.2, high Rydberg number density results in bright orange fluorescence, and low Rydberg number density in pale green fluorescence. We set up the experiment to exaggerate the laser intensity gradient (caused by absorption and divergence of the laser beams), and we show a linearised photograph of the atomic fluorescence along the 2 mm cell (Fig. 5.5). The bright orange fluorescence on the left corresponds to high Rydberg number density, and the

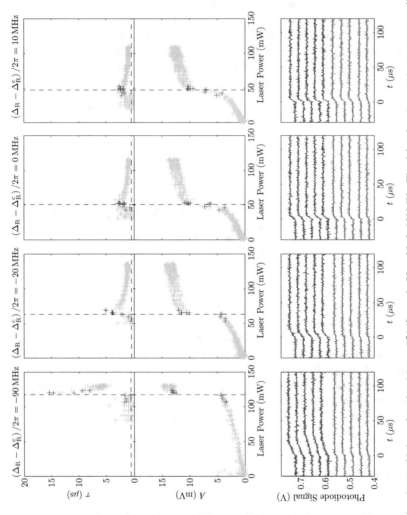

Fig. 5.4 Critical slowing down with laser detuning $(\Delta_R - \Delta_R^c)/2\pi = \{-90, -20, 0, 10\}$ MHz. The delay time (top row), response amplitude (middle row) and time-resolved transmission (bottom row) are measured when the Rydberg laser is switched on suddenly. The red lines show the fit model, $s(t)$

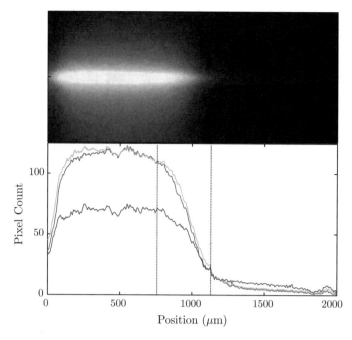

Fig. 5.5 Phase boundary. Top: Photograph showing atomic fluorescence. The laser beam propagates left to right accross the vapour cell which fills the image. However, the left side shows bright orange fluorescence (high Rydberg number density phase), where as the right hand side is dark (low Rydberg number density phase). Bottom: Cross section of the camera signal along the laser beam. We show the three camera colour channels (red, green and blue). The vertical lines, spaced 350 μm apart, indicate the 90–10% width of the domain boundary

dark region on the right corrsponds to low Rydberg number density (the pale green fluorescence is too dark to see). In contrast to the smoothly varying laser intensity (decreasing left to right), there is a sharp step in the fluorescence signal [26]. The abrupt change is a manifestation of the non-linear response, and we interpret the interface to be similar to a phase boundary separating two phases.

5.4.1 Phase Boundary Model

We extend the model outlined in Sect. 5.3 to account for spatially varying laser intensity. The cell is divided into n elements, each described by a 2-level density matrix, ρ^j, with the Rabi frequencies, Ω^j, monotonically decreasing along the chain ($j = 1 \rightarrow n$). Initially we make the naive choice to set the elements behaving independently, with each element following Eq. 5.1a. Providing the Rabi frequency varies sufficiently along the chain, the solution shows a sharp spatial transition from high to low Rydberg population. However, the transition occurs between neighbouring

Fig. 5.6 Simulations results.
We plot the excited state
population, ρ_{ee}, in a chain of
interacting elements, with
interaction length,
$\sigma = \{1, 5, 10, 20\}$. Inset: the
width of the phase boundary
(between the dashed lines) is
plotted as a function of
interaction length

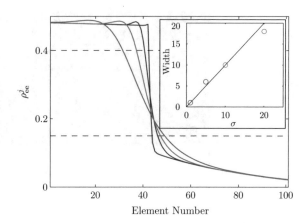

elements, and so this model does not provide any insight into the length scale of the
phase boundary.

To describe the system more closely, we now set the level shift of each element
to be influenced by its neighbours, and we write

$$W^j = V \sum_{i=1}^{n} w(i - j)\rho_{ee}^i, \qquad (5.5)$$

where V is the interaction strength, and $w(j)$ is a normalised weight function such
that $\sum_j w(j) = 1$. The coupled, non-linear differential equations,

$$\dot{\rho}_{ge}^j = i\Omega^j \left(\rho_{ee}^j - \frac{1}{2}\right) + i\left(\Delta - W^j\right)\rho_{ge}^j - \frac{\Gamma}{2}\rho_{ge}^j, \qquad (5.6)$$

$$\dot{\rho}_{ee}^j = -\Omega^j \mathrm{Im}\left(\rho_{ge}^j\right) - \Gamma\rho_{ee}^j, \qquad (5.7)$$

are solved with reflecting boundary conditions by time-stepping numerical integration until a steady state is reached. Where there is more than one steady-state (eg. the
system is bistable), the different steady-states are found by stepping the laser detuning, Δ, mimicking a laser scan. At each detuning, the system is evolved until a change
tolerance is met. We do not include any back action from the element populations
on the laser beam intensity. This simplification makes the equations much easier to
solve and can be justified by noting that the total fractional change in transmission
of the probe laser is only of order a few percent.

We show the results of the simulation in Fig. 5.6. A Gaussian weight function,
$w(j) = ke^{-j^2/\sigma^2}$, is used with $\sigma = 1, 5, 10, 20$, and all the other parameters remain
constant. As the range of the interaction increases the phase boundary becomes
wider, with the width of the boundary matching the width of the weight function.
While it is tempting to link this straight back to the experiment and deduce that

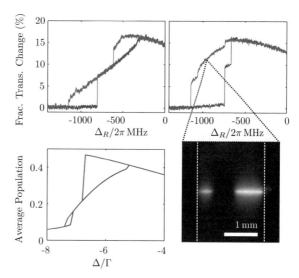

Fig. 5.7 Phase boundary connection to line shape. Left: A double hysteresis loop is seen as the Rydberg laser frequency is increasing (blue) and decreasing (red), for experiment (top) and theory (bottom). Right: Two simultaneous phase boundaries are observed. The dashed white lines in the photograph show the ends of the vapour cell

the atoms interact over a length scale matching the phase boundary width (order $100\,\mu m$), there are several other factors to consider. The severity of the laser intensity gradient of might be important, and we note that the motion of the atoms between excitation and decay might blur a sharp boundary, hiding the true distribution of the Rydberg excitation rate. However, this diffusion length might itself be relevant to the interaction responsible for the phase transition.

5.4.2 Connection to Spectral Lineshape

In addition to photographing the phase boundary, the transmission line shape was recorded as the Rydberg laser was scanned. The transmitted light relates to the whole length of the beam through the vapour, giving an integrated signal. When a phase region is created or removed, the transmission signal jumps, and when a phase boundary moves along the beam the transmission signal has a smoothly varying form. Figure 5.7 shows an example, alongside the results of a simulation carried out as outlined above, with a Gaussian weight function width $\sigma = 5$, and $n = 101$ elements. In this example the spectral line shape has a double hysteresis loop, and in the region between the two loops, a phase boundary is moving along the cell. In a separate experiment we set a laser intensity minimum inside the cell by trading focussing against absorption. On this occasion two simultaneous phase boundaries were observed.

5.5 Conclusion

Rydberg optical bistability leaves very many open questions. One of the most pressing is the search for an understanding of the microscopic mechanism that leads to the Rydberg energy level shift. In this chapter we have looked at some simple phenomenological models, but these are inherently poorly suited to addressing this question. It would be interesting to make a more complete study of the atomic fluorescence, and also to measure the presence (or absence) of ions directly. Recently in Durham similar optical bistability has been seen in a beam of strontium atoms, which would be an ideal set-up for this investigation [27]. The rest of this thesis is concerned with the interaction of Rydberg atoms with THz fields, and we will return to the phase transition in Chap. 8, in which we report the manipulation of the Rydberg phase transition with a THz field.

References

1. D. Weller, A. Urvoy, A. Rico, R. Löw, H. Kübler, Charge-induced optical bistability in thermal Rydberg vapor. Phys. Rev. A **94**, 063820 (2016)
2. C. Carr, R. Ritter, C.G. Wade, C.S. Adams, K.J. Weatherill, Nonequilibrium Phase Transition in a Dilute Rydberg Ensemble. Phys. Rev. Lett. **111**, 113901 (2013)
3. N.R. de Melo et al., Intrinsic optical bistability in a strongly driven Rydberg ensemble. Phys. Rev. A **93**, 063863 (2016)
4. D. S. Ding, C. S. Adams, B. S. Shi, G. C. Guo, *Non-equilibrium phase-transitions in multi-component Rydberg gases* (2016), arXiv:1606.08791
5. I. Kovacic, M. J. Brennan, *The Duffing Equation: Nonlinear Oscillators and Their Behaviour* (Wiley, 2011)
6. P.W. Smith, J.-P. Hermann, W.J. Tomlinson, P.J. Maloney, Optical bistability at a nonlinear interface. Appl. Phys. Lett. **35**, 846 (1979)
7. N. Kravets et al., Bistability with optical beams propagating in a reorientational medium. Phys. Rev. Lett. **113**, 023901 (2014)
8. G.I. Stegeman et al., Bistability and switching in nonlinear prism coupling. Appl. Phys. Lett. **52**, 869 (1988)
9. F.Y. Wang, G.X. Li, H.L. Tam, K.W. Cheah, S.N. Zhu, Optical bistability and multistability in one-dimensional periodic metal-dielectric photonic crystal. Appl. Phys. Lett. **92**, 211109 (2008)
10. Y.-D. Kwon, M.A. Armen, H. Mabuchi, Femtojoule-scale all-optical latching and modulation via cavity nonlinear optics. Phys. Rev. Lett. **111**, 203002 (2013)
11. S. Sarkar, J.S. Satchell, Optical bistability with small numbers of atoms. EPL (Europhysics Letters) **3**, 797 (1987)
12. C.M. Savage, H.J. Carmichael, Single atom optical bistability. IEEE J. Quantum. Electron. **24**, 1495 (1988)
13. T.E. Lee, H. Häffner, M.C. Cross, Collective quantum jumps of rydberg atoms. Phys. Rev. Lett. **108**, 023602 (2012)
14. H. Gibbs, S. McCall, T.N.C. Venkatesan, Differential gain and bistability using a sodium-filled fabry-perot interferometer. Phys. Rev. Lett. **36** (1976)
15. F.A. Hopf, C.M. Bowden, W.H. Louisell, Mirrorless optical bistability with the use of the local-field correction. Phys. Rev. A **29**, 2591 (1984)

16. R. Friedberg, S.R. Hartmann, J.T. Manassah, Mirrorless optical bistability condition. Phys. Rev. A **39**, 3444 (1989)
17. M.P. Hehlen et al., Cooperative bistability in dense, excited atomic systems. Phys. Rev. Lett. **73**, 1103 (1994)
18. C.E. Burkhardt, W.P. Garver, V.S. Kushawaha, J.J. Leventhal, Ion formation in sodium vapor containing Rydberg atoms. Phys. Rev. A **30**, 652 (1984)
19. R. Dicke, Coherence in spontaneous radiation processes. Phys. Rev. **24** (1954)
20. F. Gounand, M. Hugon, P.R. Fournier, J. Berlande, Superradiant cascading effects in rubidium Rydberg levels. J. Phys. B **12**, 547 (1979)
21. M. Reschke, Superradiance of Rydberg atoms in thermal vapour cells, Master's thesis, Universität Stuttgart (2014)
22. N. Šibalić, C.G. Wade, C.S. Adams, K.J. Weatherill, T. Pohl, Driven-dissipative many-body systems with mixed power-law interactions: Bistabilities and temperature-driven nonequilibrium phase transitions. Phys. Rev. A **94**, 011401 (2016)
23. M. Marcuzzi, E. Levi, S. Diehl, J.P. Garrahan, I. Lesanovsky, Universal nonequilibrium properties of dissipative rydberg gases. Phys. Rev. Lett. **113**, 210401 (2014)
24. R.E. Olson, Ionization cross sections for rydberg-atom-rydberg-atom collisions. Phys. Rev. Lett. **43**, 126 (1979)
25. M. Hugon, F. Gounand, P.R. Fournier, Thermal collisions between highly excited and ground-state alkali atoms. J. Phys. B Atomic Mol. Phys. **13**, L109 (1980)
26. C. Carr, Cooperative non-equilibrium dynamics in a thermal rydberg ensemble, Ph.D. thesis, Durham University(2013)
27. R. Hanley et al. In Preparation (2017)

Chapter 6
Terahertz Electrometry with Rydberg EIT

We use three-photon Rydberg electromagnetically induced transparency (EIT) to perfom Rydberg electrometry at 0.634 THz. The THz field couples the $21P_{3/2}$ and $21S_{1/2}$ caesium Rydberg states, allowing the electric field amplitude to be read-out using the optical EIT probe. A velocity-averaged, 5-level optical Bloch simulation helps us to anticipate the shape of the EIT feature in the Doppler-broadened medium, and we choose to set the Rydberg laser (799 nm) counter-propagating with respect to the probe (852 nm) and coupling (1470 nm) lasers to minimise the 3-photon Doppler shift and demonstrate a narrow Lorentzian EIT feature (8.4 ± 1.6) MHz. We show a worked example of an electric field amplitude measurement, and calibrate the internal attenuator of the THz source. The Autler-Townes lineshape is investigated when the terahertz field and Rydberg laser have both parallel and orthogonal linear polarisation.

6.1 Introduction

Rydberg atomic states possess extreme properties including long lifetimes (narrow line widths) and strong, microwave- and terahertz-frequency dipole transitions between states. These two properties make Rydberg atoms ideal platforms for probing resonant microwave and terahertz electric fields. Rydberg electrometry is the process of using the interaction between Rydberg atoms and an electric field to measure the amplitude of the electric field. Because atoms of the same isotope have identical permanent properties, Rydberg electrometry can be reliably replicated, and cross checked between species [1], making the technique suitable as a potential international measurement standard.

Rydberg microwave electrometry has seen considerable development in the last few years [2]. The technique has been used to map the spatial dependence of microwave fields [3, 4], and measure the vector electric field [5]. The effects of vapour cell geometry have been investigated, and the regime where the cell is smaller than the microwave wavelength found to be favorable [6]. Recent studies improving the

© Springer International Publishing AG, part of Springer Nature 2018   55
C. G. Wade, *Terahertz Wave Detection and Imaging with a Hot Rydberg Vapour*,
Springer Theses, https://doi.org/10.1007/978-3-319-94908-6_6

sensitivity have achieved noise equivalent power (NEP) on the order of $3\,\mu\,Vcm^{-1}\,Hz^{-1/2}$ at 5 GHz, using frequency modulation [7], Mach-Zehnder interferometery [8], and changes in the index of refraction of the vapor that result in deflection of the probe laser [9].

In this chapter we perform Rydberg electrometry in hot caesium vapour at 0.634 THz, marking the beginning of our investigation of Rydberg atoms interacting with THz fields. We give an overview of the principle of Rydberg microwave electrometry in Sect. 6.2. In Sect. 6.3 we construct a velocity-averaged, optical Bloch simulation to assess EIT line shapes in different laser beam configurations of our three-step ladder scheme. Choosing a laser configuration that gives a favorable line shape, we present a worked example of an electric field amplitude measurement and calibrate the internal attenuator of the THz source (Sect. 6.4). In Sect. 6.5 we explore the effects of the THz field polarisation on the transmission line shape.

6.2   Principle of Rydberg Microwave Electrometry

A two-level atom interacting with a resonant electric field can be described in the dressed-state picture by the Hamiltonian shown in Eq. 2.20. In the limit of strong coupling ($\Omega \gg \Gamma$), we diagonalise the Hamiltonian to give new energy eigenvalues, $E_\pm = \frac{\hbar}{2}\left(\Delta \pm \sqrt{\Delta^2 + \Omega^2}\right)$. When the electric field is on resonance ($\Delta = 0$) the system shows Autler-Townes splitting and the states $E_\pm$ have an energy interval $\hbar\Omega$. The eigenstates are superpositions of the two bare atomic states, with an electric dipole oscillating either in phase or out of phase with the electric field. If the energy interval (and hence Ω) is measured, and the electric dipole moment of the atom is known then the electric field amplitude can be deduced using Eq. 2.19. Rydberg microwave electrometry is the process of making such a measurement, using two Rydberg states as the coupled atomic energy levels. The high sensitivity of the technique originates from the dipole matrix elements between Rydberg states which can be hundreds of Debye.

In principle Autler-Townes splitting of Rydberg levels could be observed spectroscopically using a single ultra-violet laser to excite ground state atoms to the Rydberg states directly. However, this is experimentally challenging, and a more convenient method is provided by using Rydberg EIT, which gives a fast optical read-out of the Rydberg population [10]. Rydberg EIT is a non-linear effect where the transmission of a probe laser exciting a ground state to intermediate state transition is increased by the presence of a second laser driving a transition from the intermediate state to a Rydberg state. When the second laser is on resonance, it creates two competing excitation pathways with opposite amplitudes, causing the absorption of the probe field to be strongly suppressed. The probe laser is weak so that the atomic population is mostly in the ground state.

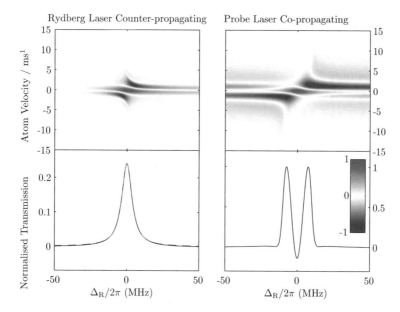

Fig. 6.1 Laser geometry effect on 3-photon EIT line shape. We show simulated transmission line shapes (bottom) and velocity class contributions (top) for the couter-propagating laser beam configuration used in this work (left) and co-propagating geometry used in previous work [11] (right). The velocity class contributions show the change of the probe laser absorption coefficient normalised to the maximum change. The bottom left plot compares the calculated line shape with a Lorentzian function FWHM 8.4 MHz (red dashed line). The laser parameters are set, $\Omega_p = \Omega_c = 2\pi \times 2\,\text{MHz}$, $\Omega_R = 2\pi \times 8.4\,\text{MHz}$

6.3 Computer Modelling

To measure Autler-Townes splitting, it is necessary to resolve two nearby EIT features, and so a simple, narrow EIT line shape is helpful. However, previous work found a triplet EIT feature in our 3-photon ladder system [11]. The origin of the complicated EIT line shape is laser absorption from moving atoms which experience a Doppler shift. In this section we use computer modelling to find a way of achieving a single, narrow Lorentzian feature by changing the relative directions of the laser beams.

The calculation is performed by generalising Eq. 2.21 to include four atomic levels and three laser driving fields, forming a ladder system to excite $21P_{3/2}$ caesium Rydberg atoms (see Sect. 3.2). We divide the vapour into a discrete set of 1D velocity classes, each of which experience a different Doppler shift. The steady state of each velocity class is found by diagonalising the time evolution operator, and selecting the time-independent eigenstate. The response of the vapour is found by summing the response of the velocity classes, weighted by a Boltzmann factor. The Rydberg laser frequency is scanned accross the atomic resonance, and we calculate the absorbtion of the probe laser beam from the coherence between the ground and first excited

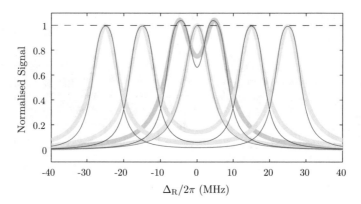

Fig. 6.2 Autler-Townes splitting simulation. We compare a 5-level optical Bloch model (coloured lines) with Lorentzian features (shaded lines), for THz Rabi frequencies, $\Omega_T = 2\pi \times \{0, 10, 30, 50\}$ MHz. The line shapes are normalised to the feature calculated for $\Omega_T = 0$. The laser parameters are set, $\Omega_p = \Omega_c = 2\pi \times 5$ MHz, $\Omega_R = 2\pi \times 8.4$ MHz

state. With the probe and coupling lasers set to zero detuning (locked to the atomic resonance), the Hamiltonian for each velocity class is given,

$$\hat{H}^{4-\text{level}} = \frac{\hbar}{2} \begin{pmatrix} 0 & \Omega_p & 0 & 0 \\ \Omega_p & 2\Delta_{1\text{ph}} & \Omega_c & 0 \\ 0 & \Omega_c & 2\Delta_{2\text{ph}} & \Omega_R \\ 0 & 0 & \Omega_R & 2\left(\Delta_R + \Delta_{3\text{ph}}\right) \end{pmatrix}, \tag{6.1}$$

where $\mathbf{v}$ is the velocity of the atoms and $\Delta_{1\text{ph}} = \mathbf{v} \cdot \mathbf{k}_p$, $\Delta_{2\text{ph}} = \mathbf{v} \cdot (\mathbf{k}_c + \mathbf{k}_c)$ and $\Delta_{3\text{ph}} = \mathbf{v} \cdot (\mathbf{k}_p + \mathbf{k}_c + \mathbf{k}_R)$. The probe, coupling and Rydberg lasers have wavevectors $k_p = 2\pi / (852\,\text{nm})$, $k_c = 2\pi / (1470\,\text{nm})$, and $k_R = 2\pi / (799\,\text{nm})$, and Rabi frequencies Ω_p, Ω_c and Ω_R respectively, and Δ_R is the detuning of the Rydberg laser. We use the condition that both the probe and coupling lasers are in the weak field limit ($I \ll I_{\text{SAT}}$).

We perform the calculation for two different coaxial laser geometries. First, we set the probe laser counter-propagating, matching the configuration used in the previous work [11]. Second, the Rydberg laser is set counter-propagating, minimising the three photon Doppler shift, $\Delta_{3\text{ph}}$, for a co-axial geometry (matching the layout shown in Fig. 3.4). Simulation results shown in Fig. 6.1 illustrate the difference between the transmission feature given by the two laser geometries. The new geometry provides a single Lorentzian feature which is narrower than the more complicated feature given by the old configuration. We adopt the new lay-out for all the Rydberg experiments described in this thesis.

The description of Autler-Townes splitting given in Sect. 6.2 would lead us to expect that a resonant microwave or terahertz field would split the Rydberg EIT feature into a doublet of Lorentzian peaks. However, the Doppler averaging in a thermal vapour makes the picture more complicated, particularly when the probe

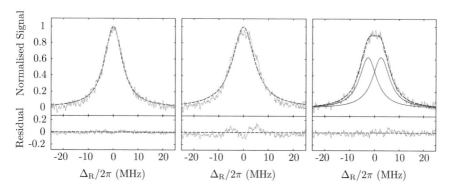

Fig. 6.3 THz electric field amplitude measurement. We show the probe transmission line shape for $\Omega_T = 0$ (left) and $\Omega_T = 2\pi \times 5.2 \pm 1.4\,$MHz (middle, right). The best fit lines (dashed) are a single Lorentzian (left, middle) and a summed pair of Lorentzian peaks (right). The dotted lines show the summed Lorentzian peaks. The Rydberg laser and THz fields have matched parallel polarisation, and the laser powers and beam waists for the probe, coupling and Rydberg lasers were measured to be $\{0.2\,\mu$W$, 40\,\mu$m$\}$, $\{0.7\,\mu$W$, 110\,\mu$m$\}$ and $\{10.2\,$mW$, 130\,\mu$m$\}$ respectively

and coupling lasers are close to saturation intensity. To investigate the Autler-Townes lineshape we add a fifth level $(21S_{1/2})$ and fourth driving field $(21P_{3/2} \rightarrow 21S_{1/2}$, $0.634\,$THz) to the optical Bloch model, and compare the result to a simple sum of two Lorentzian peaks, split by the microwave Rabi frequency (Fig. 6.2). The parameters are set so that the probe and coupling lasers are close to saturation intensity $(\Omega_{p,c}/2\pi = 5\,$MHz), matching the experimental parameters used in Sect. 6.4. In the absence of the THz field the bare EIT signal has a Lorentzian form with FWHM = $8.4\,$MHz, matching the Rydberg laser Rabi frequency. However, the doublet peaks differ substantially in shape from the simple sum of two Lorentzians, although we note that the FWHM generally matches the Lorentzian model. The doublet peak height varies, but settles to a steady value when when the Autler-Townes splitting exceeds the line width.

6.4 Electric Field Amplitude Measurement

From the considerations outlined in Sect. 6.3, we adopt a co-axial laser beam configuration with the Rydberg laser counter-propagating with respect to the probe and coupling beams. We assemble the bench layout shown in Fig. 3.4 and set the THz field resonant with the $21P_{3/2} \rightarrow 21S_{1/2}$ transition at $0.634\,$THz. The probe laser transmission is recorded using a lock-in amplifier which selects frequency components of the signal matching the modulation frequency of the Rydberg laser.

Before switching on the THz field, we measure three-photon EIT in the new laser beam configuration (Fig. 6.3, left panel). We fit a Lorentzian line shape with FWHM = $(8.4 \pm 1.6)\,$MHz, and we see from the residuals that the model is a good fit to the data. With the THz field switched on the feature broadens, and we fit two different models. To illustrate the sensitivity of the method, we start by fitting a single Lorentzian with the FWHM as a free parameter (middle panel). Despite the small change in the transmission line shape, the residuals show that the model is no longer a good fit. Instead we fit a pair of Lorentzians (right panel). Now we constrain the width of each Lorentzian peak to match the bare EIT feature, and leave the peak height and peak splitting as free parameters. We see that the model is a good fit, and we extract peak splitting corresponding to terahertz field Rabi frequency $\Omega_T/2\pi = 5.2 \pm 1.4\,\text{MHz}$. The error on the fit parameter is calculated using χ-squared analysis [12].

The THz electric field amplitude is calculated from Eq. 2.19 using the measured value of Ω_T, and a calculated dipole transition strength. We note that the THz field has linear polarisation and so we take the quantisation axis to lie along the direction of the THz electric field. In this basis, the THz field drives π transitions,

$$21P_{3/2}, m_J = -\tfrac{1}{2} \rightarrow 21S_{1/2}, m_{J'} = -\tfrac{1}{2},$$
$$21P_{3/2}, m_J = \tfrac{1}{2} \rightarrow 21S_{1/2}, m_{J'} = \tfrac{1}{2}.$$

To evaluate the transition strength, we start from the reduced dipole matrix element, $(21P_{3/2}\|e\mathbf{r}\|21S_{1/2}) = 392\,a_0 e$ [13], and we use the formalism laid out in Sect. 2.2.1 to find,

$$d = \sqrt{\frac{1}{6}}(21P_{3/2}\|e\mathbf{r}\|21S_{1/2}) = 160a_0 e. \tag{6.2}$$

Combining d and Ω_T we deduce a THz electric field amplitude of $25 \pm 5\,\text{mVcm}^{-1}$. Systematic errors originating from the calculation of the dipole matrix element can be cross checked by measuring the same electric field with separate species. Work carried out by Simons et al. [1] found a discrepancy of 0.1% in the calculated and measured sensitivity ratio between caesium and rubidium.

6.4.1 Terahertz Attenuator Calibration

We use Rydberg electrometry to characterise the voltage-controlled internal terahertz attenuator within the AMC (Fig. 6.4). As the attenuator control voltage is increased the THz field amplitude, and also the peak splitting, decrease. Although terahertz fields resulting in a THz Rabi frequency far smaller the FHWM of the EIT feature cannot be measured, there seems to be no practical limit on the maximum electric field. With this in mind, the THz beam was aligned to give the strongest possible electric field inside the vapour cell, so that the terahertz field could still be measured after the greatest possible attenuation. We measure the line shape for different attenuation voltages, and extract the terahertz Rabi frequency as a fit parameter for each measurement. From the terahertz Rabi frequency we obtain the terahertz field amplitude and the attenuation factor.

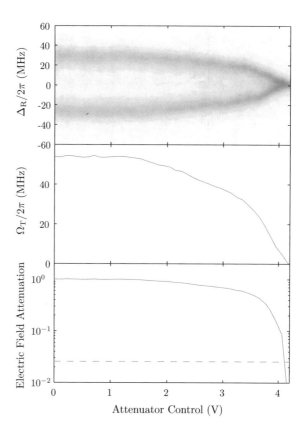

Fig. 6.4 Attenuator calibration: We present a grey-scale map showing the change of probe laser transmission in the parameter space of Rydberg laser detuning and the attenuator control voltage (top). In the middle panel we show the fitted terahertz Rabi frequency, Ω_T. The relative attenuation is shown in the bottom panel, and the measurement floor is indicated with a dashed line (derived using the 1.4 MHz errorbar found in Sect. 6.4)

Once the fractional change of the THz field amplitude has been measured, the attenuator can be used as a lever to infer THz fields that are too small to be measured directly. We set up the system so that the unattenuated terahertz beam results in the smallest measurable terahertz electric field amplitude inside the cell. When we then attenuate the THz beam we multiply the measured field by the attenuation factor to obtain the new electric field amplitude. The calibrated range of the attenuator is over 20 dB.

6.5 Polarisation Effects

When the THz and Rydberg laser have crossed polarisation, the laser excites atoms to Rydberg m_j states that are not coupled by the THz field. Taking the THz electric field direction as a quantisation axis once again, the laser field (perpendicular to the quantisation axis) drives a coherent sum of σ^+ and σ^- transitions and excites

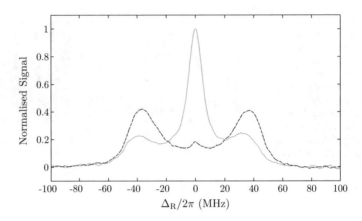

Fig. 6.5 Laser polarisation effect on line shape. We show the probe transmission when the Rydberg laser polarisation is parallel (dashed) and perpendicular (full) to the THz electric field. The laser powers and beam waists for the probe, coupling and Rydberg lasers were $\{0.3\,\mu W, 50\,\mu m\}$, $\{1\,\mu W, 50\,\mu m\}$ and $\{4\,mW, 60\,\mu m\}$ respectively

$21P_{3/2}$, $m_j = \pm 3/2$ states. However, the THz field only drives π-transitions, leaving the $21P_{3/2}$, $m_j = \pm 3/2$ states uncoupled.

Figure 6.5 shows the result of laser polarisation on the transmission line shape. Keeping the THz field polarisation constant, we show the line shape with the laser polarised parallel (blue) and perpendicular (red) to the THz field. When the fields are orthogonal we see a large central peak corresponding to the states that are uncoupled from the THz field, and the Autler-Townes splitting of the coupled states only adds a pair of smaller satellite peaks. When the electric fields of the laser beam and terahertz field are parallel, the $21P_{3/2}m_j = \pm 3/2$ states are not excited, and so all the Rydberg atoms should contribute to the Autler-Townes signal. However, in the data we see a small residual peak that probably comes from either a slight polarisation mis-match, or a region of the laser beam path where the terahertz field is absent.

Finally we note that the width of the doublet peaks is now very much greater than the width of the bare EIT feature in Fig. 6.3. As we will see in the next chapter, the broadening is due to spatial variation of the terahertz field.

6.6 Conclusion

In this section we have presented Rydberg electrometry at 0.634 THz, the highest frequency reported to date. Because terahertz-frequency atomic transitions are generally weaker than equivalent microwave transitions, we do not expect to match the sensitivity achieved in the microwave frequency band [7, 8]. The technique also falls behind the sensitivity of established techniques in the terahertz band [14], however it is the traceability to SI units and the application for calibrating terahertz fields the lend Rydberg electrometry its utility. Using the internal attenuator of our terahertz source we can measure electric field amplitudes over a range of $\approx 20\,\text{dB}$, and we will go on to use our measurements to calibrate the sensitivity of Rydberg terahertz imaging (Chap. 7), and to investigate a terahertz-driven phase transition (Chap. 8).

References

1. M.T. Simons, J.A. Gordon, C.L. Holloway, Simultaneous use of Cs and Rb Rydberg atoms for dipole moment assessment and RF electric field measurements via electromagnetically induced transparency. J. Appl. Phys. **120**, 123103 (2016)
2. J.A. Sedlacek et al., Microwave electrometry with Rydberg atoms in a vapour cell using bright atomic resonances. Nat. Phys. **8**, 819 (2012)
3. H.Q. Fan, S. Kumar, R. Daschner, H. Kübler, J.P. Shaffer, Subwavelength microwave electric-field imaging using Rydberg atoms inside atomic vapor cells. Opt. Lett. **39**, 3030 (2014)
4. C.L. Holloway et al., Sub-wavelength imaging and field mapping via electromagnetically induced transparency and autler-townes splitting in rydberg atoms. Appl. Phys. Lett. **104**, 244102 (2014)
5. J. Sedlacek, A. Schwettmann, H. Kübler, J.P. Shaffer, Atom-based vector microwave electrometry using rubidium rydberg atoms in a vapor cell. Phys. Rev. Lett. **111**, 063001 (2013)
6. H. Fan et al., Effect of vapor-cell geometry on rydberg-atom-based measurements of radio-frequency electric fields. Phys. Rev. Appl. **4**, 044015 (2015)
7. S. Kumar, H. Fan, H.Kübler, A.J. Jahangiri, J.P. Shaffer, Rydberg-atom based radio-frequency electrometry using frequency modulation spectroscopy in room temperature vapor cells. Opt. Express, **25**, 8625 (2017)
8. S. Kumar, H. Fan, H.Kübler, J. Sheng, J.P. Shaffer, Atom-based sensing of weak radio frequency electric fields using homodyne readout. Sci. Rep. **7**, 42981 (2017)
9. H. Fan, S. Kumar, H. Kbler, J.P. Shaffer, Dispersive radio frequency electrometry using rydberg atoms in a prism-shaped atomic vapor cell. J. Phys. B **49**, 104004 (2016)
10. A.K. Mohapatra, T.R. Jackson, C.S. Adams, Coherent optical detection of highly excited rydberg states using electromagnetically induced transparency. Phys. Rev. Lett. **98**, 113003 (2007)
11. C. Carr et al., Three-photon electromagnetically induced transparency using rydberg states. Opt. Lett. **37**, 3858 (2012)
12. I. Hughes, T. Hase, *Measurements and Their Uncertainties* (Oxford University Press, 2010)
13. N. Šibalić, J. Pritchard, C. Adams, K. Weatherill, ARC: an open-source library for calculating properties of alkali Rydberg atoms. Comput. Phys. Commun. **220** 319–331 (2017)
14. A. Rogalski, F. Sizov, Terahertz detectors and focal plane arrays. Opto-Electronics Rev. **19**, 346 (2012)

Chapter 7
Real-Time Near-Field Terahertz Field Imaging

Terahertz induced Rydberg atomic fluorescence is photographed to make an image of a Terahertz standing wave. We measure the fluorescence spectrum and the sensitivity bandwidth, and investigate the (sub-wavelength) resolution limit set by atomic motion. The camera signal is calibrated using Rydberg electrometry. Consecutive frames from a 25 Hz video are presented, demonstrating the real-time capability of this technique.

7.1 Introduction

Terahertz (THz) near-field imaging is a flourishing discipline [1, 2], with applications from the characterisation of metameterials [3, 4] and waveguides [5, 6] to fundamental studies of beam propagation [7]. Sub-wavelength imaging typically involves rastering structures or detectors with length scale shorter than the radiation wavelength; in the THz domain this has been achieved using a number of techniques including apertures [8] and scattering tips [9, 10]. Alternatively, mapping THz or microwave fields onto an optical wavelength and imaging the visible light removes the requirement for scanning a local probe, speeding up image collection times [11, 12]. Atomic states that couple to multiple transitions offer an interface between wavelengths, and in this way atomic ground states have been used to image microwave fields with an optical probe [13]. While atomic ground states are only sensitive to a limited selection of microwave frequencies, Rydberg atoms couple to strong, electric dipole transitions across a wide range of microwave and THz frequencies. However, previous methods for THz imaging with Rydberg atoms did not use a wavelength conversion strategy, but the THz radiation was used to ionise the Rydberg

This chapter includes work published:
'Real-time near-field terahertz imaging with atomic optical fluorescence',
C. G. Wade, N. Šibalić, J. M Kondo, N. R. de Melo C. S. Adams, and K. J. Weatherill,
Nature Photonics **11** 40-43 (2017).

atoms instead. The ions were then accelerated towards a screen where they formed an image [14, 15].

In this chapter we photograph terahertz-induced optical fluorescence from Rydberg atoms to image THz fields. Section 7.2 outlines the characteristics of the atomic fluorescence. We present a standing wave image in Sect. 7.3, and discuss background fluorescence and blurring from atomic motion. In Sect. 7.4, we present 25 frames per second near-field imaging.

7.2   Fluorescence Signal

We excite Rydberg atoms to the $21S_{1/2}$ state using the excitation scheme shown in Fig. 3.2. The bench layout described in Fig. 3.4 is repeated, and we ensure that the intensity and detuning of the Rydberg laser is set so that the optical bistability described in Chap. 5 is not present. The Rydberg laser and terahertz fields are both set off-resonant from their respective atomic transitions, but in such a way that $\Delta_R = \Delta_T$. This Raman condition means that laser excitation of the $21S_{1/2}$ state is energetically allowed, but direct excitation of the $21P_{3/2}$ state is suppressed. As a result Rydberg atoms are only created in places where the THz field and laser beams overlap in space, and so the optical fluorescence is localised to areas where the THz field is present. A camera records an image of the optical atomic fluorescence, allowing us to infer the spatial distribution of the THz field.

7.2.1   Fluorescence Spectrum

Even in the absence of optical bistability, the atomic fluorescence includes emission lines from very many Rydberg states, both above and below the energy of the $21S_{1/2}$ state. Figure 7.1 shows an example of the fluorescence spectrum. Red light originates from Rydberg atoms in nP and nF states decaying to the 5D state; green spectral lines encompass nS and nD atoms decaying to the $6P_{3/2}$ and $6P_{1/2}$ states, with the strongest contributions coming from nD states; blue light originates from atoms decaying from the 7P to 6S ground state, the last step in a cascade process. The complexity of the fluorescence spectrum originates from state transfer in the Rydberg manifold, and may be linked to the creation of ions and the level shift responsible for optical bistability (Sect. 5.1.1). However, the fast Rydberg population transfer processes are likely to accelerate the optical decay which would benefit the spatial resolution of the THz images (see Sect. 7.3.3).

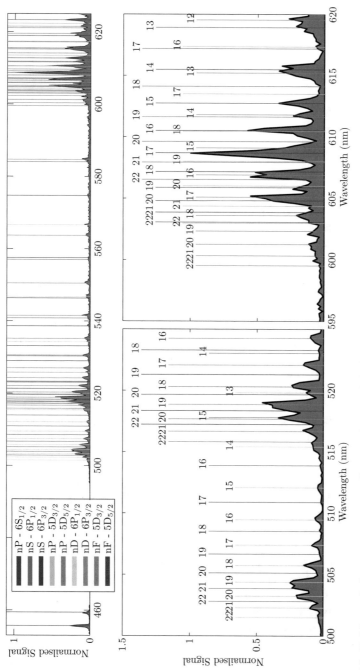

Fig. 7.1 Visible atomic fluorescence spectrum. Top: complete scan from 450–630 nm. Bottom: Selected regions of the spectrum are shown with labelled atomic transitions. The number associated with each line refers to the principle quantum number of the decaying Rydberg state, and the colour indicates the angular momentum of the Rydberg and final states. The data were recorded with $\Delta_R = \Delta_T = 2\pi \times -259\,\text{MHz}$ and the cell temperature was 71°C

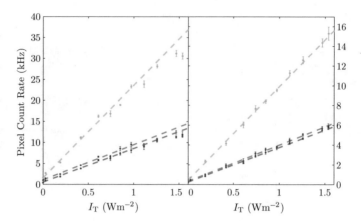

Fig. 7.2 Camera Calibration: We present the camera signal as a function of THz intensity for $\Delta_R/2\pi = \Delta_T/2\pi = -135$ MHz (left) and $\Delta_R/2\pi = \Delta_T/2\pi = -243$ MHz (right). The red, green and blue (rgb) data points and fit lines correspond to the rgb channels in the camera respectively. The error bars represent statistical variation in five repeated measurements

7.2.2   Image Sensitivity Calibration

For absolute calibration of the camera signal we use the Rydberg EIT technique to make a measurement of the THz electric field amplitude along the laser beams (as explained in Chap. 6). The camera has three color channels (red, green and blue), which are each sensitive to a different selection of optical transitions (unfortunately *Canon* were unwilling to divulge the spectral sensitivity of the camera). We measure the signal in each channel as a function of THz field intensity, I_T (Fig. 7.2). The camera ISO is set to 25600, and the exposure time is adjusted so that the 14-bit CCD does not saturate. The 'pixel count rate' is then the CCD output divided by the exposure time, though we note that this does not correspond to photon counting.

The system response is set by the detuning, $\Delta_R = \Delta_T$, and we show calibration curves for $\Delta_T/2\pi = \{-135, -243\}$ MHz. The sensitivity when $\Delta_T/2\pi = -135$ MHz (22.1 ± 0.3 kHzW^{-1}m^2, green channel) is greater than when $\Delta_T/2\pi = -243$ MHz (9.5 ± 0.1 kHzW^{-1}m^2, green channel, first six points only). However, this is at the cost of increased background signal, and saturation of the signal when the THz intensity $I_T > 1.2$ Wm^{-2}. In both cases the blue and red channels are less sensitive than the green channel. The sensitivity is also influenced by the laser intensity and cell temperature. For the images presented in this chapter we set $\Delta_T/2\pi = -243$ MHz and the cell temperature to 68°C. The probe, coupling and Rydberg lasers have $1/e^2$ beam radii and power $\{30\,\mu\text{m}, 21\,\mu\text{W}\}$, $\{90\,\mu\text{m}, 89\,\mu\text{W}\}$ and $\{130\,\mu\text{m}, 520\,\text{mW}\}$ respectively.

Fig. 7.3 Spectral
bandwidth: The red, green
and blue camera channel
sensitivity (red, green and
blue points respectively) is
shown as a function of the
terahertz field detuning, Δ_T.
The Rydberg laser detuning
is set $\Delta_R/2\pi = -243$ MHz
(shown by the dashed line),
and the THz intensity is set
at 1 Wm^{-2}.

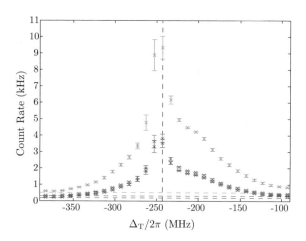

7.2.3 Sensitivity Bandwidth

Resonant Rydberg atomic transitions give the fluorescence imaging a narrow-band
sensitivity. To characterise the sensitivity bandwidth we set the Rydberg laser detuning $\Delta_R/2\pi = -243$ MHz, and step the THz field frequency across the Raman
condition ($\Delta_T = \Delta_R$). The camera signal for the three colour channels is shown in
Fig. 7.3. The width of the response is around 100 MHz, but we note an interesting
asymmetry with a shoulder clearly present on the high frequency side. The author is
unable to explain this asymmetry without invoking the effects of inter-atomic interactions. It is possible to speculate that the shape of the feature indicates a distribution
of environments experienced by Rydberg atoms within the vapour. The width of the
narrow central peak is dominated by uncertainty in the Rydberg laser detuning.

7.3 Standing Wave Image

We present a camera image of a THz standing wave (Fig. 7.4). The features in the
image can be understood by considering the geometry of the laser and THz fields:
both the laser and THz beams propagate horizontally across the image, but part of
the THz beam is reflected back on itself to produce a standing wave interference
pattern with nodes and anti-nodes perpendicular to the laser beam. The fluorescence
has the form of a horizontal line matching the width of the laser beams, with periodic
modulation according to the THz intensity of the standing wave. As expected from
the signal calibration the fluorescence mostly appears green.

The lower panel of Fig. 7.4 shows the signal cross section for both the photograph
in the upper panel and a photograph taken when the THz field was absent (background
fluorescence). In each case we show the pixel count rate averaged between the dashed

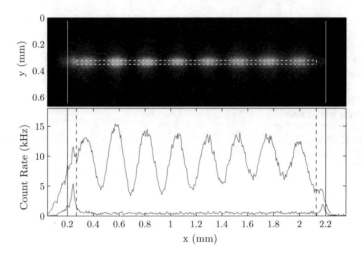

Fig. 7.4 Standing Wave Image. Top: THz-induced atomic fluorescence (proportional to the THz intensity) is imaged. The dark vertical stripes indicate the nodes of the THz standing wave, spaced at $\lambda/2$ intervals (237 μm). Bottom: Cross section of the THz standing wave image (red) and background fluorescence (blue). The dashed lines show the region used to model the line shape, and the bold lines show the extent of the 2 mm vapour cell

lines. The 2 mm vapour cell occupies the region $0.2 < x/\text{mm} < 2.2$, and we see that the background fluorescence shows a peak at each end of the vapour cell (see Sect. 7.3.2).

7.3.1 Autler-Townes Line Shape Model

The frequency of the THz field is adjusted to match the atomic resonance and we use Rydberg EIT to measure an Autler-Townes line shape corresponding to the standing wave image (Fig. 7.5). Spatial variation of the THz field amplitude leads to broadening of the line shape. In order to cross check the image with the transmission measurement, we make a self consistent model for the line shape, using the distribution of THz field intensities from the image.

To construct a model for the line shape, it is first necessary to interpret the distribution of THz field strengths from the image. Because the camera signal is proportional to the *intensity*, but the Autler-Townes splitting is proportional to the *amplitude*, we write

$$E(x) = E_0\sqrt{F(x) - F_0(x)} \Big/ \Big\langle \sqrt{F(x) - F_0(x)} \Big\rangle, \qquad (7.1)$$

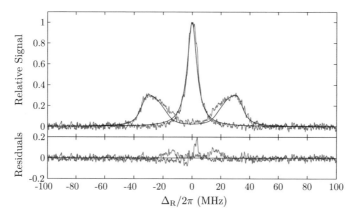

Fig. 7.5 THz electric field measurement using electromagnetically induced transparency (EIT): Rydberg EIT is measured for the standing wave image (red), and with the THz field switched off (blue). The peak splitting is proportional to the THz electric field amplitude, and so the doublet peaks are broadened by the field variation in the standing wave

where $E(x)$ is the THz field amplitude, $F(x)$ is the fluorescence intensity, F_0 is the background fluorescence intensity and $\langle \ldots \rangle$ denotes an average value along the laser beams.

The laser beam is considered in n segments as it passes through the vapour, each with length δx corresponding to the size of one camera pixel. After n segments, the laser intensity exiting the vapour, I, is given,

$$I = I_0 \prod_{i=1}^{n} e^{-\alpha_i \delta x}, \tag{7.2}$$

where I_0 is the laser intensity entering the vapour. The absorption coefficient of the ith segment, α_i, is dominated by background absorption (α_0), but is modified slightly by the THz field, so we write,

$$\alpha_i = \alpha_0 - g_i(E, \Delta), \tag{7.3}$$

where $g_i(E, \Delta)$ is a function of the local THz electric field amplitude, E, and the Rydberg laser detuning, $\Delta/2\pi = f - f_0$, where f_0 is the resonant transition frequency. The form of $g_i(E, \Delta)$ is given by Autler-Townes splitting of the EIT feature,

$$g_i(E, \Delta) = \frac{A_0}{2} \left(\frac{\left(\frac{\Gamma}{2}\right)^2}{\left(\Delta - \frac{\Omega}{2}\right)^2 + \left(\frac{\Gamma}{2}\right)^2} + \frac{\left(\frac{\Gamma}{2}\right)^2}{\left(\Delta + \frac{\Omega}{2}\right)^2 + \left(\frac{\Gamma}{2}\right)^2} \right), \tag{7.4}$$

where Γ is the Full Width Half Maximum (FWHM) of of the doublet peaks and Ω is the Rabi frequency of the THz transition, related to the electric amplitude field as

$E = \hbar\Omega/d$ with d as the dipole matrix element. Our investigation in Sect. 6.3 shows that the height of the doublet peaks is independent of the splitting (provided the splitting is well resolved), and the width, Γ, matches the bare EIT feature. Therefore, although A_0 is a fit parameter, it is held constant for all n elements.

We assume $\sum_{i=1}^{n} g_i \delta x \ll 1$ and write from Eq. 7.2,

$$I(\Delta) = I'\left(1 + \delta x \sum_{i=1}^{n} g_i\right), \tag{7.5}$$

where $I' = I_0 \prod_{i=1}^{n} e^{-\alpha_0 \delta x}$ is the background intensity. We fit the lineshape $I(\Delta)$ using parameters A_0, E_0 and f_0, but we constrain the value of Γ to match the EIT feature when the THz field is absent (Fig. 7.5). To remove bias from the background fluorescence at the edges of the cell we exclude these regions (vertical dashed lines in Fig. 7.4). The pixel intensities are binned to speed up the fitting routine.

Although the model closely resembles the data, we note that the line width of the data is still broader than the model. This could be due to reduced fringe visibility in the image caused by blurring from atomic motion, or non-uniform sensitivity along the laser path length as the laser intensity changes through the cell.

7.3.2   Cell Wall Interaction

While discussing Fig. 7.4 we noted that the background fluorescence shows a peak at each end of the vapour cell. We show the results of a systematic study of this effect in Fig. 7.6. Each photograph shows the background fluorescence for different laser detuning. For $\Delta_R < 0$ there is consistently a bright green area at each end of the cell (left column), however, the bright area is absent for $\Delta_R > 0$ (right column). Instead, the fluorescence is suppressed at the edge.

It might be the case that the behaviour is caused by stray charge on the surface of the vapour cell. The action of a local electric field would be a negative shift of the $21P_{3/2}$ Rydberg state, causing it to come into resonance with a negatively detuned laser beam, but further away from resonance for a positively detuned laser beam. It might also be relevant that the ends of the vapour cell correspond to the edge of a cooperative medium. Atoms near the surface have fewer neighbours with which to interact, and so the possibility for a collectively sustained excitation is reduced. It would be interesting to explore whether the edges of the cell play a role in nucleating the phases of the Rydberg vapour described in Sect. 5.4.

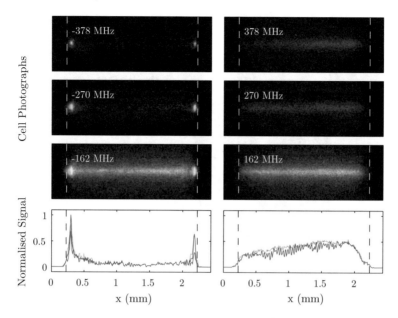

Fig. 7.6 Cell wall interaction. Top: Images of background fluorescence for labeled Rydberg laser detuning, $\Delta_R/2\pi$. Bottom: Cross section for $\Delta_R/2\pi$ = -162 MHz and $\Delta_R/2\pi$ = 162 MHz. The extent of the vapour cell is marked by dashed lines

7.3.3 Motional Blurring

Motion of the atoms in the time between excitation and decay results in slight blurring of the features in the image. However this motion-induced blurring is minimised in the x-direction (along the laser beams) by Doppler selection, which suppresses the excitation of fast moving atoms. The Doppler shift changes the laser frequency experienced by moving atoms according to the scalar product of the atom velocity, v, and the vector sum, k, of the laser wave-vectors. Atoms moving in the x-direction parallel to k experience mis-matched laser frequencies which inhibit laser excitation, suppressing the contribution to the signal from fast moving atoms. It is due to Doppler selection that we are able to see the interference fringes in the camera images.

Each fluorescence line has a characteristic decay time and consequently each colour channel (red, green, blue) of the camera shows a different resolution (Fig. 7.7). We infer that the red channel corresponds to the slowest decay processes as it shows the smallest intensity fringe visibility, $V = 30\%$ (extracted from a model for the THz intensity, $I = \left[1 - V\cos(2kx + \phi)\right](ax + b)$, with k the wave-vector of the THz wave, and parameters a, b and ϕ fitted to the data). The green and blue channels both show higher intensity fringe visibility (45%) due to faster decay times. The fringe contrast implies 24% reflection of the THz electric field amplitude. The spatial

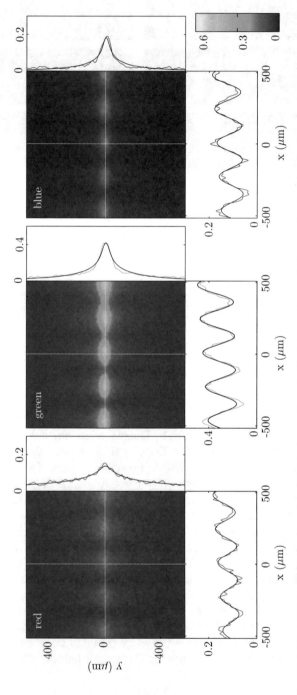

Fig. 7.7 Motional blurring: Each colour channel of the image shown in Fig. 7.4 records a particular set of optical decay pathways, giving different resolution according to the decay times. Here we present the red, green and blue channels separately and show x and y cross sections of the camera channels at the positions indicated by white dotted lines

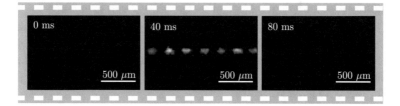

Fig. 7.8 Real time imaging: Sequence of frames from a 25 frame per second video. The THz field is gated to be on and off in consecutive frames

resolution could be optimized by using a narrow-band spectral filter to choose the fastest decay processes. Motion blurring seems a sensible candidate to consider for the mis-match between the data and line shape model in Fig. 7.5.

7.4 Real-Time Imaging

The image acquisition time is set by the fluorescence intensity, the camera sensitivity and the desired signal to noise ratio. Using a generic consumer digital camera we record 25 frames per second video. In Fig. 7.8 we show three consecutive video frames during which the THz source was gated in synchrony with the camera shutter so that alternating frames saw the field present or absent. In the second frame the standing wave is visible, while the first and third frames are blank. This video-rate imaging without averaging over repeated experimental runs establishes the real-time character of the imaging. The underlying limit of the imaging bandwidth is likely set by the fluorescence lifetime of the Rydberg atoms which is on the order of microseconds.

7.5 Conclusion

In this chapter we have demonstrated THz to optical conversion using an atomic vapour and presented real-time, near-field THz field images. Our method presents a significant benefit over THz field mapping where photoconductive antennae [5] or electro-optic crystals [7, 16] are rastered, leading to image acquisition on the time scale of tens of hours. Recent high-resolution work imaging THz surface waves still required around a 1 hour collection time despite claiming a significant speed-up [17]. The caesium vapour does not absorb the THz field, and so we are able to image narrowband THz waves traveling in opposite directions superposed as a standing wave. This would be impossible using either detectors that absorb energy from the THz field such as pixel arrays [18], or detectors that rely on short, high-intensity THz pulses. Because the THz-to-optical conversion medium is in the gas phase we envisage immersing structures within the caesium vapour in order to measure diffraction and reflection around wavelength-scale obstructions. The linear response of the atomic fluorescence relies on relatively weak Rydberg excitation, which is achieved by setting the Rydberg laser and terahertz fields away from the atomic resonance. In the next chapter we consider the response of a resonantly excited Rydberg vapour to terahertz fields.

References

1. A.J.L. Adam, Review of near-field terahertz measurement methods and their applications. J. Infrared Millim. Terahertz Waves **32**, 976 (2011)
2. W.L. Chan, J. Deibel, D.M. Mittleman, Imaging with terahertz radiation. Rep. Prog. Phys. **70**, 1325 (2007)
3. A. Bitzer et al., Terahertz near-field imaging of electric and magnetic resonances of a planar metamaterial Abstract: Opt. Express **17**, 1351 (2009)
4. G. Acuna et al., Surface plasmons in terahertz metamaterials. Opt. Express **16**, 18745 (2008)
5. O. Mitrofanov, T. Tan, P.R. Mark, B. Bowden, J.A. Harrington, Waveguide mode imaging and dispersion analysis with terahertz near-field microscopy. Appl. Phys. Lett. **94**, 171104 (2009)
6. K. Nielsen et al., Bendable, low-loss Topas fibers for the terahertz frequency range. Opt. Express **17**, 8592 (2009)
7. A. Bitzer, M. Walther, Terahertz near-field imaging of metallic subwavelength holes and hole arrays. Appl. Phys. Lett. **92**, 231101 (2008)
8. A.J. Baragwanath et al., Terahertz near-field imaging using subwavelength plasmonic apertures and a quantum cascade laser source. Opt. Lett. **36**, 2393 (2011)
9. P. Dean et al., Apertureless near-field terahertz imaging using the self-mixing effect in a quantum cascade laser. Appl. Phys. Lett. **108**, 091113 (2016)
10. A.J. Huber, F. Keilmann, J. Wittborn, J. Aizpurua, R. Hillenbrand, Terahertz Near-Field Nanoscopy of Nanodevices. Nano Lett. **8**, 3766 (2008)
11. Q. Wu, T.D. Hewitt, X.-C. Zhang, Two-dimensional electro-optic imaging of THz beams. Appl. Phys. Lett. **69**, 1026 (1996)
12. A. Doi, F. Blanchard, H. Hirori, K. Tanaka, Near-field THz imaging of free induction decay from a tyrosine crystal. Opt. Express **18**, 1161 (2010)
13. A. Horsley, G.-X. Du, P. Treutlein, Widefield microwave imaging in alkali vapor cells with sub-100 µm resolution. New J. Phys. **17**, 112002 (2015)
14. M. Drabbels, L.D. Noordam, Infrared imaging camera based on a Rydberg atom photodetector. Appl. Phys. Lett. **74**, 1797 (1999)
15. A. Gurtler, A.S. Meijer, W.J. van der Zande, Imaging of terahertz radiation using a Rydberg atom photocathode. Appl. Phys. Lett. **83**, 222 (2003)
16. M.A. Seo et al., Fourier-transform terahertz near-field imaging of one-dimensional slit arrays: mapping of vectors. Opt. Express **15**, 11781 (2007)
17. X. Wang et al., Visualization of terahertz surface waves propagation on metal foils. Sci. Rep. **6**, 18768 (2016)
18. A.W. Lee, Q. Hu, Real-time, continuous-wave terahertz imaging by use of a microbolometer focal-plane array. Opt Lett **30**, 2563 (2005)

Chapter 8
Terahertz-Driven Phase Transition in a Hot Rydberg Vapour

We use a weak terahertz-frequency field ($I_T \ll 1\,\mathrm{Wm}^{-2}$) to drive a non-equilibrium phase transition in a hot caesium Rydberg vapour. We measure a phase diagram in the parameter space of terahertz field and laser detuning, and we find a linear shift of the critical laser detuning with coefficient $-179 \pm 2\,\mathrm{MHzW}^{-1}\mathrm{m}^{-2}$. Considering the system as a terahertz detector we calculate sensitivity, $S \approx 90\,\mu\mathrm{Wm}^2\mathrm{Hz}^{-1/2}$. The phase transition is accompanied by hysteresis and bistability, allowing the system to act as a latch controlled by a 1 ms terahertz pulse with energy of order 10 fJ.

8.1 Introduction

Phase transitions are widely studied for both fundamental inquiry and application driven research, from data storage [1] to shape memory alloys [2], however examples of phase transitions driven by terahertz fields are limited. Controlling magnetic domains with terahertz-frequency magnetic fields could open opportunities for ultrafast data storage, but while terahertz magnetisation dynamics have been observed using strong fields ($B > 0.1\,\mathrm{T}$) [3, 4], permanent magnetic switching has so far been largely elusive [5]. Terahertz-frequency electric fields have been used to drive the metal-insulator transition in VO_2 [6, 7], but again a strong field was required ($E > 1\,\mathrm{MVcm}^{-1}$). The reversal of a superconducting phase transition through heating from low-intensity terahertz radiation has been exploited for transition edge sensors (TES) [8], but requires cryogenic temperatures.

In this chapter we demonstrate a THz-driven phase transition in a hot Rydberg vapour. The phase transition between low- and high Rydberg number density can be driven above room temperature using a weak terahertz-frequency field,

This chapter includes work accepted for publication:
'A terahertz-driven non-equilibrium phase transition in a room temperature atomic vapour',
C. G. Wade, M. Marcuzzi, E. Levi, J. M. Kondo, I. Lesanovsky C. S. Adams and K. J. Weatherill.

© Springer International Publishing AG, part of Springer Nature 2018
C. G. Wade, *Terahertz Wave Detection and Imaging with a Hot Rydberg Vapour*,
Springer Theses, https://doi.org/10.1007/978-3-319-94908-6_8

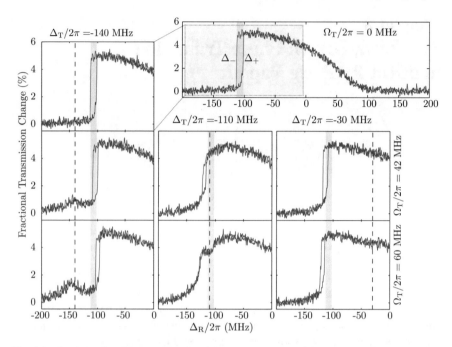

Fig. 8.1 Bistability line shape modification: The grey shaded area highlights the bistable scan range before the THz field is applied. The THz field has detuning $\Delta_T/2\pi =$-140 MHz (left), -110 MHz (middle) and -30 MHz (right), shown by the dashed lines. The THz field has Rabi frequency $\Omega_T/2\pi = \{0, 42, 60\}$ MHz (top, middle and bottom)

$I_T \ll 1\,\mathrm{Wm}^{-2}$ ($E_T \ll 1\,\mathrm{Vcm}^{-1}$). The behaviour forms an extension of the optical bistability observed in Chap. 5. In Sect. 8.2 we measure a phase diagram, and consider the application of the system as a terahertz detector. Hysteresis in the system response allows the demonstration of a latch, whereby the system is permanently altered by a transient terahertz field (Sect. 8.3.2). Transitions between non-equilibrium states, such as the collective driven-dissipative Rydberg phases discussed here, are also of particular physical interest because they require a theoretical framework beyond the well established thermodynamics of equilibrium systems.

8.2 Spectral Line Modification

We report the modification of Rydberg optical bistability by a resonant THz field. In Chap. 5 we saw that optical bistability leaves a signature of hysteresis, and so we begin our investigation by examining how the hysteresis changes when the terahertz field is introduced.

The level scheme and bench layout described in Chap. 3 are repeated, and we use the Rydberg laser to excite atoms to the $21P_{3/2}$ state. The THz field is generated by an amplifier multiplier chain (AMC) and drives the $21P_{3/2}$ to $21S_{1/2}$ transition which is resonant at 0.634 THz. The transmission of the probe beam is recorded as the Rydberg laser is scanned, and all other parameters are held constant, with the probe and coupling laser frequencies locked to the atomic resonances. Before the addition of the terahertz field, the laser power and cell temperature are set to produce a hysteresis signal free from phase boundaries (Sect. 5.4.1). The low-to-high and high-to-low Rydberg density phase transitions occur at Rydberg laser detuning $\Delta_R = \Delta_+$ and $\Delta_R = \Delta_-$ respectively, making the system bistable for $\Delta_- < \Delta_R < \Delta_+$. In this instance, $\Delta_-/2\pi = -110$ MHz and $\Delta_+/2\pi = -100$ MHz.

When the THz field is introduced, the hysteresis window shifts to a new range, $\Delta'_- < \Delta_R < \Delta'_+$, and we define shift parameters, $\delta_- = \Delta'_- - \Delta_-$ and $\delta_+ = \Delta'_+ - \Delta_+$. The changes depend on the THz field Rabi frequency, Ω_T, and detuning, Δ_T, and we present experimental measurements for a selection of parameters in Fig. 8.1. We identify three different regimes:

(i) $\Delta_T < \Delta_-$ **(left column)**: The addition of the THz field shifts the hysteresis window shifts such that $\delta_- > 0$. Furthermore the hysteresis window narrows, with $\delta_- > \delta_+ > 0$. Finally, an extra peak is visible at the 4-photon Raman resonance, $\Delta_R = \Delta_T$.

(ii) $\Delta_T > \Delta_+$ **(right column)**: The hysteresis window shifts once again, but now in the opposite direction, $\delta_+, \delta_- < 0$. However, the hysteresis window still narrows, and we see that $\delta_+ > \delta_-$ The peak at the 4-photon resonance is now almost hidden, but there is still a small bump visible in the data.

(iii) $\Delta_T \approx \Delta_-, \Delta_+$ **(middle column)**: The hysteresis is initially shifted (middle row), but closes completely at higher power (bottom row).

The data in Fig. 8.1 are a subset of a larger data set, mapping the parameter space in Δ_R, Δ_T and Ω_T. The data were collected using a labview program to automate the data collection, allowing fast repetitions of the Rydberg laser scan whilst stepping the THz frequency. We collate the results with $\Omega_T/2\pi = 60$ MHz as a phase map (Fig. 8.2), similar to the optical bistability phase map shown in Sect. 5.3.1. Where the response is monostable we shade the phase map red, and where the response is bistable we shade it blue. We see that the system is bistable in two separate regions of the parameter space.

We can begin to explain the phase map by considering the dressed states, $|+\rangle$ and $|-\rangle$ with energy $E_\pm = \frac{\hbar}{2}\left(\Delta_T \pm \sqrt{\Delta_T^2 + \Omega_T^2}\right)$ respectively. In case (i), state $|+\rangle$ has mostly $21P_{3/2}$ character, and has a positive AC Stark shift $E_+ > 0$, which is mirrored by the bistable region, resulting in $\delta_-, \delta+ > 0$. State $|-\rangle$ is apparent as the new peak in the spectrum. In case (ii), state $|-\rangle$ has mostly $21P_{3/2}$, but has a negative AC Stark shift $E_- < 0$. Again the shift is followed by the bistable region, this time resulting in $\delta_-, \delta+ < 0$ In case (iii) the hysteresis closes because the four-photon resonance provides rapid Rydberg excitation regardless of the system's history. It is hoped that these phenomena can be more satisfactorily explained by a non-linear

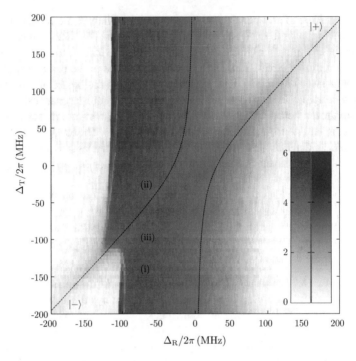

Fig. 8.2 Terahertz bistability phase map ($\Omega_T/2\pi = 60\,$MHz). In monostable regions (red) the colourmap shows the fractional transmission change in the probe beam (%). In bistable regions (blue) we show the difference signal. The black lines show the dressed states, $|+\rangle$ and $|-\rangle$. The cell temperature was 77 °C and the probe, coupling and Rydberg lasers had $1/e^2$ radii and beam power $\{30\,\mu\text{m}, 70\,\mu\text{W}\}$, $\{90\,\mu\text{m}, 28\,\mu\text{W}\}$ and $\{130\,\mu\text{m}, 310\,\text{mW}\}$ respectively

optical Bloch simulation formulated by *E. Levi, M. Marcuzzi* and *I. Lesanovsky* [9]. The model involves extending the 2-level scheme described in Sect. 5.3 to include a third level, representing the atomic state coupled by the terahertz field. However, the development is a work-in-progress and the details are yet to be finalised.

8.2.1   Linear Shift of Hysteresis Laser Detuning

We make a systematic study of the frequency shift of the hysteresis loop up to Terahertz intensity $I_T = 30\,\text{mWm}^{-2}$. The THz detuning is set $\Delta_T = -90.7\,$MHz, such that $\Delta_- < \Delta_T < 0$. The Rydberg laser is scanned using the AOM for speed, and the transmission of the probe beam is measured. For each shift measurement the laser frequency is cycled twice: once to measure the hysteresis loop whilst the THz field is blocked before the cell, and again to measure the response with the THz field admitted into the cell. By comparing the two measurements we read off δ_+ and δ_-,

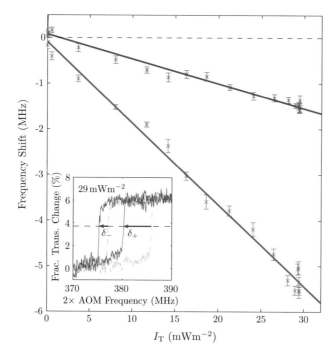

Fig. 8.3 Phase Transition Shift. We show δ_+ (red) and δ_- (blue) as a function of THz intensity, with a straight line fit. Inset: δ_+ and δ_- are read from a frequency scan controlled by an AOM. The bold (faint dashed) line shows the hysteresis with the THz field at intensity $I_T = 29\,\text{mWm}^{-2}$ ($I_T = 0\,\text{mWm}^{-2}$), whilst the laser frequency is increased (red) and decreased (blue). The THz intensity is measured using the Rydberg electrometry technique presented in Chap. 6

although the absolute values Δ'_-, Δ'_+ are not known. The differential measurement helps to cancel changes in Δ_+ and Δ_- caused by fluctuating laser power, and the complete pair of cycles takes 2 ms.

We show the results of the experiment in Fig. 8.3. For each value of Ω_T the measurement was repeated 10 or 11 times, and we plot the mean and standard error of the shifts. The standard error in the mean has typical value $\sigma \approx 0.1$ MHz. We find a linear dependence of the hysteresis shift on the THz field intensity, I_T, and fit a straight line $\delta = m I_T + c$ with the coefficients found using linear regression:

	$m/\text{MHzW}^{-1}\text{m}^2$	c/MHz
δ_+	-179 ± 2	-0.09 ± 0.04
δ_-	-54 ± 1	0.07 ± 0.03

We consider using the hysteresis shift as a way to measure the THz field intensity. Taking δ_+, we estimate the noise equivalent power, NEP $= \sqrt{t\sigma^2/m^2} \approx 90\,\mu\text{Wm}^{-2}\text{Hz}^{-1/2}$, where t is the time required to perform the ten pairs of cy-

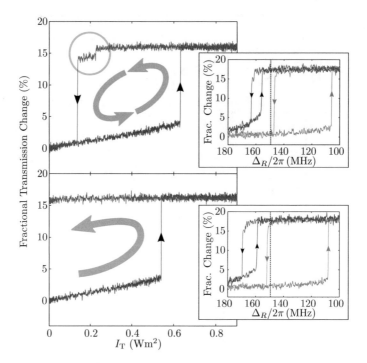

Fig. 8.4 THz Bistability: The THz intensity is cycled when $\Delta_R < \Delta_-$ (top) and $\Delta_- < \Delta_R < \Delta_+$ (bottom). The green ring highlights a step in the phase transition, corresponding to the creation of a phase boundary. Inset: The bold [faint] line shows the response as the laser frequency is increased (red) and decreased (blue) with THz intensity $I_T = 0.9\,\mathrm{Wm}^{-2}$ [$I_T = 0\,\mathrm{mWm}^{-2}$]. The dashed line shows the detuning of the Rydberg laser for the main plot

cles. The sensitivity compares unfavourably to microwave Rydberg EIT electrometry where $\mathrm{NEP} = 3\,\mu\mathrm{Vcm}^{-1}\mathrm{Hz}^{-1/2}$ has been achieved at 5GHz (equivalent to $0.1\,\mathrm{nWm}^{-2}\mathrm{Hz}^{-1/2}$) [10]. This is partly due to the smaller dipole moment associated with the terahertz-frequency transition, but is also because the non-linear response of the terahertz phase transition precludes the use of lock-in amplification for noise reduction. Nevertheless, the work here only represents an initial study, and further optimisation could lead to enhanced performance.

8.3 Static Response to Terahertz Intensity

We study the system response to changes in the THz intensity while all the other system parameters are held constant. From the phase map shown in Fig. 8.2 we expect the response to be dictated by Δ_T and Δ_R in relation to Δ_- and Δ_+. We limit ourselves to setting the Terahertz detuning, $\Delta_T > \Delta_-$, such that $\delta_+, \delta_- < 0$. The

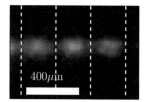  

Fig. 8.5 Terahertz phase domain photographs. Left: A back reflection of the THz beam causes a standing wave pattern in the atomic fluorescence. The nodes are spaced slightly further than half a wavelength apart (237 μm), because the reflected beam is not coaxial with the incident beam. Middle, Right: A boundary between Rydberg high- and low-density Rydberg collective states is pinned between nodes of the standing wave. The interface is located at the boundary between the bright orange and pale green fluorescence

THz intensity is cycled from $I_T = 0 \rightarrow 0.9\,\text{Wm}^{-2}$ and back $I_T = 0.9 \rightarrow 0\,\text{Wm}^{-2}$ over a period of 300 ms, and we record the transmission of the probe laser. The cell temperature was 71°C and the probe, coupling and Rydberg lasers had $1/e^2$ radii and beam power $\{65\,\mu\text{m}, 37\,\mu\text{W}\}$, $\{50\,\mu\text{m}, 56\,\mu\text{W}\}$ and $\{65\,\mu\text{m}, 260\,\text{mW}\}$ respectively. The results for two different regimes are presented in Fig. 8.4:

(i) $\mathbf{\Delta_R < \Delta_-}$ **(top row)**: At the beginning of the cycle ($I_T = 0$) the system is monostable, but as the THz intensity increases it becomes bistable, fulfilling the condition $\Delta'_- < \Delta_R < \Delta'_+$. However, at the maximum intensity ($I_T = 0.9\,\text{Wm}^{-2}$) the system is once again monostable, with $\Delta_R > \Delta'_+$. The bistable response is traced out by the system as a hysteresis loop, and we see two abrupt changes which we identify as Rydberg phase transitions.

(ii) $\mathbf{\Delta_- < \Delta_R < \Delta_+}$ **(bottom row)**: At the beginning of the cycle ($I_T = 0$) the system is now bistable. Once again we see the phase transition as the intensity is increased, making the system monostable once $\Delta_R > \Delta'_+$. However, as I_T decreases again, the system does not switch back to its original phase. Note that through both cycles Δ_R remains constant.

In general the shape of the hysteresis loop can be tuned by changing Δ_R, Δ_T and Ω_R. The open hysteresis of case (ii) forms the basis for a latching detector (Sect. 8.3.2).

8.3.1 Phase Domain Hopping

The abrupt change in probe laser transmission associated with the terahertz phase transition is sometimes broken into several smaller steps such as the one highlighted in Fig. 8.4. As we saw in Sect. 5.4, steps in the probe laser transmission can correspond to spatial structure along the length of the laser beam. To investigate terahertz-induced phase boundaries we create a THz standing wave inside the vapour cell and set the laser beam intensity decreasing through the cell.

In Fig. 8.5 we show a series of photographs of the atomic fluorescence. The terahertz standing wave is visible as modulation of green atomic fluorescence in the low rydberg density phase (see Chap. 7), and we also see the bright orange fluorescence

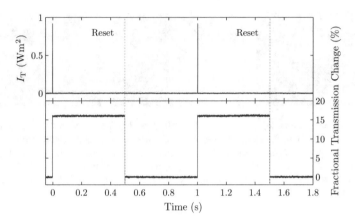

Fig. 8.6 Latching detector: A 1 ms pulse from the THz source flips the system between bistable phases. The system remains in its altered state until the Rydberg laser power is cycled (dashed line, 'reset')

corresponding to the high Rydberg density phase. A phase boundary is present between the bright orange fluorescence (left), and pale green fluorescence (right), and we see that it is pinned to the nodes of the standing wave as the terahertz field is perturbed.

8.3.2 Latching Detector

We describe a protocol for the system to act as a latching detector. The system is set up for open hysteresis (Fig. 8.4), and we initialise the vapour in the lower branch of the hysteresis loop, with the terahertz field switched off (blue trace, Fig. 8.4). Next the THz field is switched on for 1 ms (the shortest time achievable with our equipment), with suffcient intensity to drive the phase transition. In this instance the necessary intensity was $0.55\,\mathrm{Wm^{-2}}$ and pulse intensity was $1\,\mathrm{Wm^{-2}}$. Once the pulse is complete the system remains in the upper branch of the hysteresis loop, corresponding to a significant Rydberg population, which is sustained by the laser excitation and Rydberg level shift (Sect. 5.1.1). The system remains in the upper branch of the hysteresis loop until the laser parameters drift, or the system is reset by turning the Rydberg laser off and on again (Fig. 8.6).

In a similar way to the critical slowing down measured in Sect. 5.3.2, the switching time becomes slow when the system parameters only pass the phase transition by a small margin. Switching times between $10\,\mu\mathrm{s}$ and $100\,\mu\mathrm{s}$ were observed, with the fastest times achieved when the THz intensity of the pulse far exceeded the minimum intensity. Taking $10\,\mu\mathrm{s}$ switching with $1\,\mathrm{Wm^{-2}}$ and a $50\,\mu\mathrm{m}$ waist implies that the system is sensitve to THz pulses on the order of 10 fJ. The latching Rydberg detector compares favourably with recent work demonstrating a photoacoustic detector sensitive to individual 3.6 nJ pulses at a 500 Hz repetition rate [11].

8.4 Conclusion

In this chapter we have seen how the Rydberg atom phase transition can be manipulated by THz radiation. The work constitutes an example of a THz-driven phase transition and might have strong applications for use in THz detection. The nonlinear response means that the performance is determined by the chosen protocol, and the work here is only initial groundwork. For digital communication it is possible to conceive an extension to the latching protocol whereby the bistable system is completely controlled by the THz field without the need for the laser reset. In the scheme a pulse with $\Delta_T > \Delta_+$ would push the system from the low to high transmission state (as shown in Sect. 8.3.2), and a pulse with $\Delta_T < \Delta_-$ would perform the opposite operation. After each pulse the system would 'latch' into the state in which it was put.

Beyond the exciting practical applications, there are a lot of unanswered physical questions about the system. It is possible that future work using the THz field might illuminate the Rydberg shift discussed in chapter 5. It would also be interesting to investigate THz bistability at frequencies corresponding to different atomic transitions (Fig. 1.2). The data presented here used the $21P_{3/2} \rightarrow 21S_{1/2}$ transition, but the $26P_{3/2} \rightarrow 26D_{5/2}$ and $24P_{3/2} \rightarrow 26D_{5/2}$ transitions also yielded similar results.

References

1. S.S. Parkin, M. Hayashi, L. Thomas, Magnetic domain-wall racetrack memory. Science **320**, 190 (2008)
2. J.M. Jani, M. Leary, A. Subic, M.A. Gibson, A review of shape memory alloy research, applications and opportunities. Mater. Des. **56**, 1078 (2014)
3. C. Vicario et al., Off-resonant magnetization dynamics phase-locked to an intense phase-stable terahertz transient. Nat. Photonics **7**, 720 (2013)
4. T. Kampfrath et al., Coherent terahertz control of antiferromagnetic spin waves. Nat. Photonics **5**, 31 (2011)
5. M. Shalaby, C. Vicario, C.P. Hauri, Simultaneous electronic and the magnetic excitation of a ferromagnet by intense THz pulses. New J. Phys. **18**, 013019 (2016)
6. M. Liu et al., Terahertz-field-induced insulator-to-metal transition in vanadium dioxide metamaterial. Nature **487**, 345 (2012)
7. Z.J. Thompson et al., Terahertz-triggered phase transition and hysteresis narrowing in a nanoantenna patterned vanadium dioxide film. Nano Lett. **15**, 5893 (2015)
8. W. Zhang et al., *High sensitive THz superconducting hot electron bolometer mixers and transition edge sensors*, in *SPIE/COS Photonics Asia*, pp. 1003009–1003009, International Society for Optics and Photonics (2016)
9. C.G. Wade et al., *A terahertz-driven non-equilibrium phase transition in a room temperature atomic vapour*, arXiv:1709.00262
10. S. Kumar, H. Fan, H. Kübler, A.J. Jahangiri, J.P. Shaffer, *Rydberg-atom based radio-frequency electrometry using frequency modulation spectroscopy in room temperature vapor cells*. Opt. Express **25**, 8625 (2017)
11. S.-L. Chen et al., Efficient real-time detection of terahertz pulse radiation based on photoacoustic conversion by carbon nanotube nanocomposite. Nat. Photonics **8**, 537 (2014)

Chapter 9
Summary and Outlook

In this thesis we have studied and exploited the interaction between terahertz frequency electric fields and laser excited Rydberg atoms in a hot caesium vapour. We have used caesium Rydberg atoms to perform terahertz electrometry and real-time terahertz field imaging and we have explored a terahertz-driven non-equilibrium phase transition in a Rydberg vapour. In developing the necessary experimental techniques we have investigated hyperfine quantum beats modified by an excited-state transition, and Rydberg optical bistability.

In the future the work presented here could be extended in many different directions. The mechanism for interaction between Rydberg atoms responsible for optical bistability still remains an open question. Using the monochromator to monitor the fluorescence of particular atomic decay channels as the Rydberg laser is scanned might reveal some more insight, particularly at low principal quantum number, n, where the fluorescence is strong and it is possible to resolve individual lines. In contrast, using a longer vapour cell might help investigate higher Rydberg states ($n \gg 30$). The stronger interactions of higher lying states require a lower number density, and so a longer vapour cell could be used to retain sufficient optical depth.

Throughout this thesis we focused on one particular terahertz-frequency transition (caesium $21P_{3/2} \rightarrow 21S_{1/2}$), but there are very many others to investigate. There are also several other options for immediately developing the terahertz imaging technique. Although the laser beam configuration used is this work leads to a convenient Rydberg EIT signal, it gives the minimum 3-photon Doppler shift for atoms moving along coaxial laser beams. Because the Doppler shift is responsible for velocity selection which reduces the motional blurring, it is reasonable to imagine that better resolution could be achieved with a different laser beam configuration, (all the lasers co-propagating would maximise the 3-photon Doppler shift). So far the imaging technique has only been demonstrated in 1-dimension, and so another immediate development could be to use the laser beams in a light-sheet configuration. Velocity selection could be achieved in both axes by crossing the excitation laser beams. The author hopes that this work represents the beginning of further terahertz-field detection and imaging with hot Rydberg Vapours.

© Springer International Publishing AG, part of Springer Nature 2018

C. G. Wade, *Terahertz Wave Detection and Imaging with a Hot Rydberg Vapour*,
Springer Theses, https://doi.org/10.1007/978-3-319-94908-6_9

Publications Related to this Thesis

- *'Probing an excited-state atomic transition using hyperfine quantum-beat spectroscopy'*, **C. G. Wade**, N. Šibalić, J. Keaveney, C. S. Adams, and K. J. Weatherill, Phys. Rev. A, **90**, 033424 (2014)
- *'Intrinsic optical bistability in a strongly driven Rydberg ensemble'*, N. R. de Melo, **C. G. Wade**, N. Šibalić, J. M. Kondo, C. S. Adams, and K. J. Weatherill, Phys. Rev. A, **93**, 063863 (2016)
- *'Real-time near-field terahertz imaging with atomic optical fluorescence'*, **C. G. Wade**, N. Šibalić, J. M Kondo, N. R. de Melo C. S. Adams, and K. J. Weatherill, Nature Photonics **11** 40-43 (2017)
- *'A terahertz-driven non-equilibrium phase transition in a room temperature atomic vapour'*, **C. G. Wade**, M. Marcuzzi, E. Levi, J. M. Kondo, I. Lesanovsky C. S. Adams and K. J. Weatherill, (Accepted for publication in Nature Communications)

© Springer International Publishing AG, part of Springer Nature 2018
C. G. Wade, *Terahertz Wave Detection and Imaging with a Hot Rydberg Vapour*,
Springer Theses, https://doi.org/10.1007/978-3-319-94908-6

About the Author

Chris Wade grew up in the UK, attending St. Bartholomew's School, Newbury, for his secondary education, before studying Natural Sciences at Fitzwilliam College, University of Cambridge where he obtained MSci (Hons) *Experimental and Theoretical Physics (*1st class) in 2012. His masters thesis supervised by Dr. Bill Allison, '*Contrast Mechanisms in Scanning Helium Microscopy*', was shortlisted for the UK 'Science, Engineering and Technology (SET) Award'.

In 2012 he started his Ph.D studies at Durham University under the supervision of Dr. Kevin Weatherill and Prof. Charles Adams, during which time the work presented in this thesis was undertaken alongside teaching undergraduate students and engaging with outreach activities. Since completing his Ph.D, Chris Wade has taken up a post-doctoral research position at the University of Oxford with Prof. Ian Walmsley, investigating quantum metrology within the UK's Networked Quantum Information Technology (NQIT) hub.

© Springer International Publishing AG, part of Springer Nature 2018   91
C. G. Wade, *Terahertz Wave Detection and Imaging with a Hot Rydberg Vapour*,
Springer Theses, https://doi.org/10.1007/978-3-319-94908-6

Printed in the United States
By Bookmasters